Solution Manual

for

Mechanics and Control of Robots
(Springer, 1997)

by

K. C. Gupta

ISBN 978-1-4612-7308-0 ISBN 978-1-4612-1840-1 (eBook)
DOI 10.1007/978-1-4612-1840-1

Preface

This *Solution Manual* contains solutions to the problems at the end of each chapter in *Mechanics and Control of Robots*, Springer, 1997. These problems range in complexity from simple to difficult. The instructors using this book as a main text or a supplemental text will find it useful to preview the solutions before assigning the corresponding problems for homework.

The problems in the book belong to one of the following three categories:

(a) Practice Problems -- These relate most directly to the material presented in the text and are designed to give the students some practice with the concepts or equations presented in the book.

(b) Extensions -- These problems will allow the students to explore the text material further. Hints are included either separately, or within the problem statement itself.

(c) Supplementary Material -- As explained in the preface of the book, the level of the book is kept at an introductory level appropriate for first year graduate students in engineering. Some additional material of difficult or tedious nature is included within some of the problems. Proper hints are provided to guide the students.

As of this writing, two errors in the book are as follows:

Page 42, Problem 15, line 4: .... related as $\Delta R = [\varepsilon]\, R_1$ and ......

Page 143, Figure 4.2(a): The third axis (on the right) should be $\mathbf{u}_{k+1}$

Any other errors in the book or the solution manual can be communicated directly to the author.

K. C. Gupta
July, 1997

CHAPTER 1

1. (a)

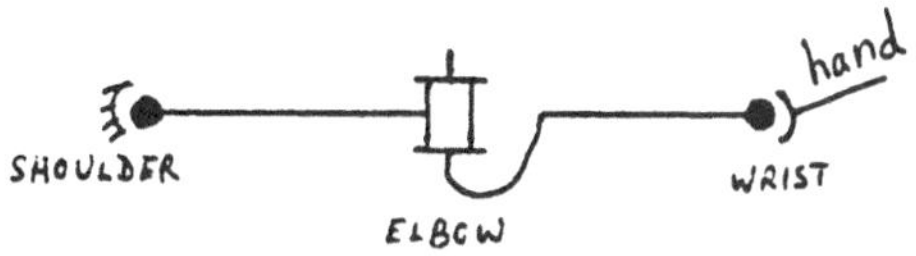

$\text{dof} = \sum f_i = 3+1+3 = 7$

(b)

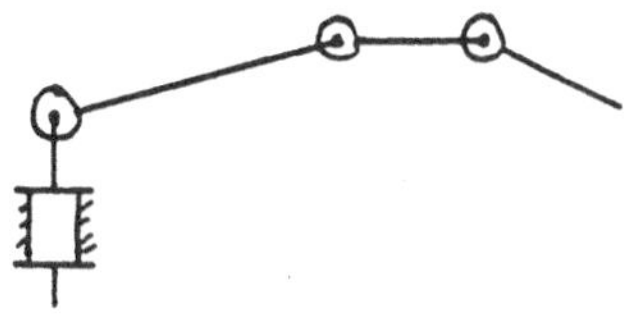

$\text{dof} = 4\times 1 = 4 \quad (= \sum f_i)$

2. (a)

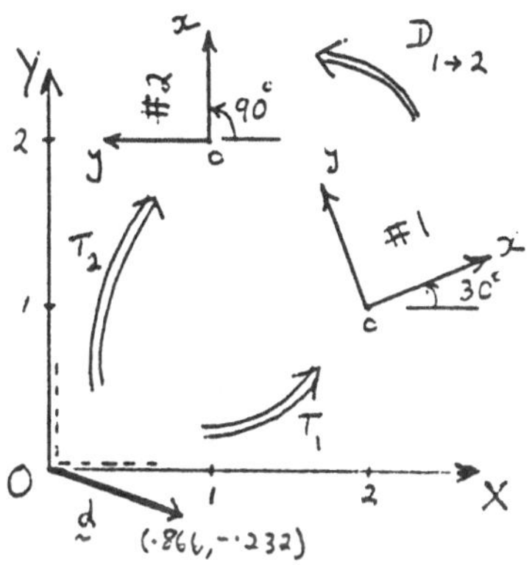

$$\begin{bmatrix} X_1 \\ Y_1 \\ 1 \end{bmatrix} = T_1 \begin{bmatrix} x \\ y \\ 1 \end{bmatrix} = \begin{bmatrix} .866 & -.5 & 2 \\ .5 & .866 & 1 \\ 0 & 0 & 1 \end{bmatrix} \begin{bmatrix} x \\ y \\ 1 \end{bmatrix}$$

$$\begin{bmatrix} X_2 \\ Y_2 \\ 1 \end{bmatrix} = T_2 \begin{bmatrix} x \\ y \\ 1 \end{bmatrix} = \begin{bmatrix} 0 & -1 & 1 \\ 1 & 0 & 2 \\ 0 & 0 & 1 \end{bmatrix} \begin{bmatrix} x \\ y \\ 1 \end{bmatrix}$$

Last column of T_i contains (X_{0i}, Y_{0i})

(b) Eliminate (x, y) to get the direct relation between (X_1, Y_1) & (X_2, Y_2)

$$\begin{bmatrix} X_2 \\ Y_2 \\ 1 \end{bmatrix} = T_2 T_1^{-1} \begin{bmatrix} X_1 \\ Y_1 \\ 1 \end{bmatrix} = \begin{bmatrix} .5 & -.866 & .866 \\ .866 & .5 & -.232 \\ 0 & 0 & 1 \end{bmatrix} \begin{bmatrix} X_1 \\ Y_1 \\ 1 \end{bmatrix} \equiv D_{1\to 2} \begin{bmatrix} X_1 \\ Y_1 \\ 1 \end{bmatrix}$$

Last column of $D_{1\to 2}$ contains $\underset{\sim}{d}$: $(.866, -.232)$

(c) $D_{1\to 2}$: Body is rotated by 60° and the body point coincident with the base origin O is displaced by $\underset{\sim}{d}$: $(.866, -.232)$

See the sketch for "displacements" T_1, T_2 & $D_{1\to 2}$.

3. (a) $X_i = x' \cos\theta_i - y' \sin\theta_i$

$Y_i = x' \sin\theta_i + y' \cos\theta_i$

$Z_i = z' + s_i$

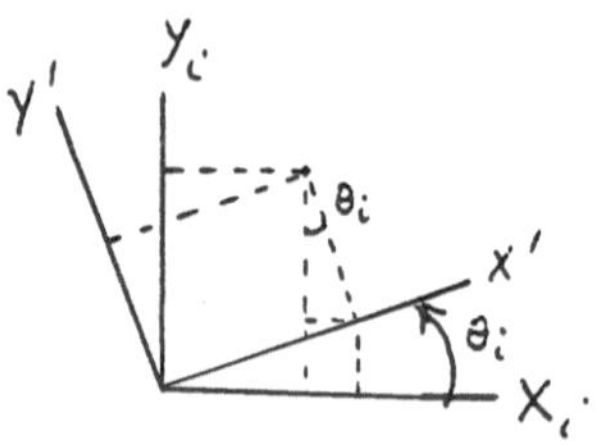

(b) $x' = X_{i+1} + a_i$

$y' = Y_{i+1} \cos\alpha_i - Z_{i+1} \sin\alpha_i$

$z' = Y_{i+1} \sin\alpha_i + Z_{i+1} \cos\alpha_i$

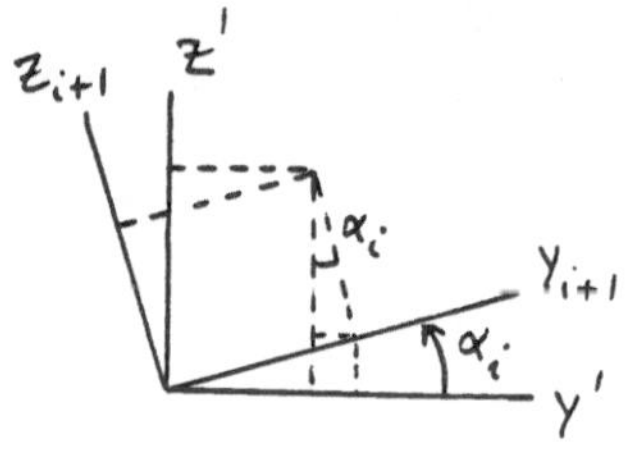

(c) After some algebra,

$$\left(\frac{\partial A_i}{\partial \theta_i}\right) A_i^{-1} = \begin{bmatrix} 0 & -1 & 0 & 0 \\ 1 & 0 & 0 & 0 \\ 0 & 0 & 0 & 0 \\ 0 & 0 & 0 & 0 \end{bmatrix} \quad \& \quad \left(\frac{\partial A_i}{\partial s_i}\right) A_i^{-1} = \begin{bmatrix} 0 & 0 & 0 & 0 \\ 0 & 0 & 0 & 0 \\ 0 & 0 & 0 & 1 \\ 0 & 0 & 0 & 0 \end{bmatrix}$$

4.

Variables $\theta_i; s_i$

Fixed $\theta_{j,0}$; $s_{j,0}$

Choices X_1, X_2, X_4, X_7 (location)

choices X_3, X_5, X_6 (direction)

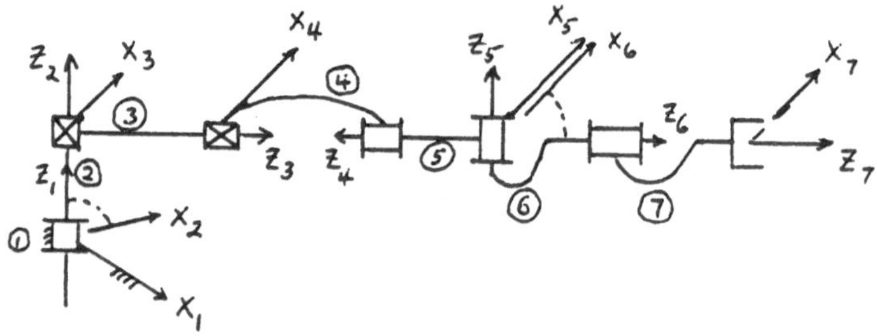

k	α_k	a_k	θ_k	s_k
1	0	0	θ_1	0
2	+90°	0	$\theta_{2,0}$	s_2
3	180°	0	0	s_3
4	+90°	0	θ_4	$s_{4,0}$
5	+90°	0	θ_5	0
6	0	0	θ_6	$s_{6,0}$

$$A_1 = \begin{bmatrix} c\theta_1 & -s\theta_1 & 0 & 0 \\ s\theta_1 & c\theta_1 & 0 & 0 \\ 0 & 0 & 1 & 0 \\ 0 & 0 & 0 & 1 \end{bmatrix} \quad A_2 = \begin{bmatrix} c\theta_{2,0} & 0 & s\theta_{2,0} & 0 \\ s\theta_{2,0} & 0 & -c\theta_{2,0} & 0 \\ 0 & 1 & 0 & s_2 \\ 0 & 0 & 0 & 1 \end{bmatrix} \quad A_3 = \begin{bmatrix} 1 & 0 & 0 & 0 \\ 0 & -1 & 0 & 0 \\ 0 & 0 & -1 & s_3 \\ 0 & 0 & 0 & 1 \end{bmatrix}$$

$$A_4 = \begin{bmatrix} c\theta_4 & 0 & s\theta_4 & 0 \\ s\theta_4 & 0 & -c\theta_4 & 0 \\ 0 & 1 & 0 & s_{4,0} \\ 0 & 0 & 0 & 1 \end{bmatrix} \quad A_5 = \begin{bmatrix} c\theta_5 & 0 & s\theta_5 & 0 \\ s\theta_5 & 0 & -c\theta_5 & 0 \\ 0 & 1 & 0 & 0 \\ 0 & 0 & 0 & 1 \end{bmatrix} \quad A_6 = \begin{bmatrix} c\theta_6 & -s\theta_6 & 0 & 0 \\ s\theta_6 & c\theta_6 & 0 & 0 \\ 0 & 0 & 1 & s_{6,0} \\ 0 & 0 & 0 & 1 \end{bmatrix}$$

5.

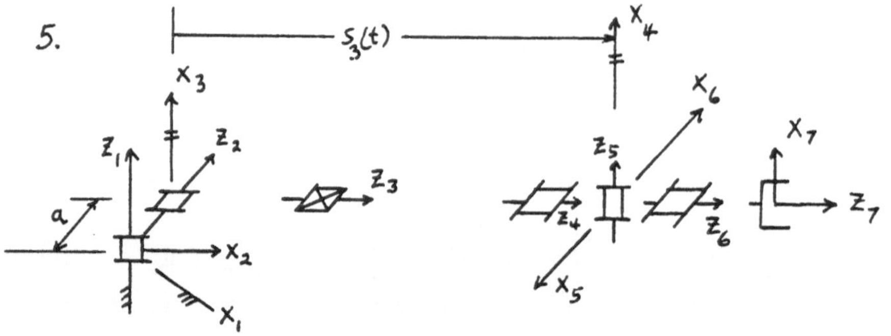

i	θ_i	S_i	α_i	a_i
1	$\theta_1(t)$	0	$-90°$	0
2	$\theta_2(t)$	a	$-90°$	0
3	0	$S_3(t)$	0	0
4	$\theta_4(t)$	0	$90°$	0
5	$\theta_5(t)$	0	$90°$	0
6	$\theta_6(t)$	h	0	0

For the position shown,

$\theta_2 = -90°$, $\theta_4 = 90°$,

$\theta_5 = 180°$, $\theta_6 = 90°$

$$A_1 = \begin{bmatrix} c\theta_1 & 0 & -s\theta_1 & 0 \\ s\theta_1 & 0 & c\theta_1 & 0 \\ 0 & -1 & 0 & 0 \\ 0 & 0 & 0 & 1 \end{bmatrix}; \quad A_2 = \begin{bmatrix} c\theta_2 & 0 & -s\theta_2 & 0 \\ s\theta_2 & 0 & c\theta_2 & 0 \\ 0 & -1 & 0 & a \\ 0 & 0 & 0 & 1 \end{bmatrix}; \quad A_3 = \begin{bmatrix} 1 & 0 & 0 & 0 \\ 0 & 1 & 0 & 0 \\ 0 & 0 & 1 & S_3 \\ 0 & 0 & 0 & 1 \end{bmatrix}$$

$$A_4 = \begin{bmatrix} c\theta_4 & 0 & s\theta_4 & 0 \\ s\theta_4 & 0 & -c\theta_4 & 0 \\ 0 & 1 & 0 & 0 \\ 0 & 0 & 0 & 1 \end{bmatrix}; \quad A_5 = \begin{bmatrix} c\theta_5 & 0 & s\theta_5 & 0 \\ s\theta_5 & 0 & -c\theta_5 & 0 \\ 0 & 1 & 0 & 0 \\ 0 & 0 & 0 & 1 \end{bmatrix}; \quad A_6 = \begin{bmatrix} c\theta_6 & -s\theta_6 & 0 & 0 \\ s\theta_6 & c\theta_6 & 0 & 0 \\ 0 & 0 & 1 & h \\ 0 & 0 & 0 & 1 \end{bmatrix}$$

6.

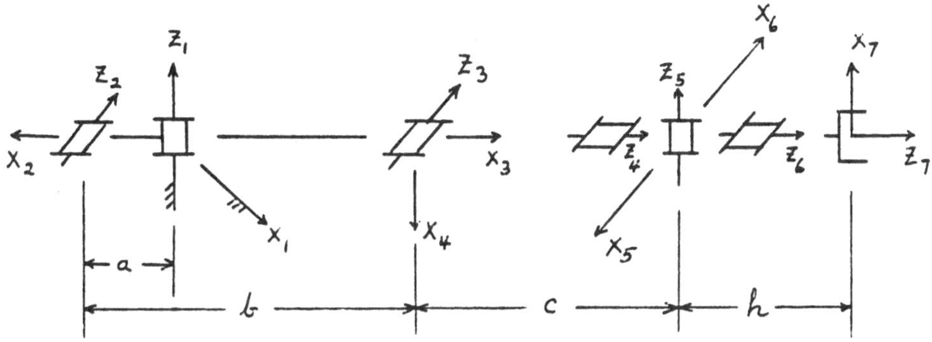

i	θ_i	s_i	α_i	a_i
1	$\theta_1(t)$	0	90°	a
2	$\theta_2(t)$	0	0	b
3	$\theta_3(t)$	0	90°	c
4	$\theta_4(t)$	c	90°	0
5	$\theta_5(t)$	0	90°	0
6	$\theta_6(t)$	h	0	0

For the position shown,

$\theta_2 = 180^\circ$, $\theta_3 = 90^\circ$, $\theta_4 = -90^\circ$

$\theta_5 = 180^\circ$, $\theta_6 = 90^\circ$

$$A_1 = \begin{bmatrix} c\theta_1 & 0 & s\theta_1 & a\,c\theta_1 \\ s\theta_1 & 0 & -c\theta_1 & a\,s\theta_1 \\ 0 & 1 & 0 & 0 \\ 0 & 0 & 0 & 1 \end{bmatrix}; \quad A_2 = \begin{bmatrix} c\theta_2 & -s\theta_2 & 0 & b\,c\theta_2 \\ s\theta_2 & c\theta_2 & 0 & b\,s\theta_2 \\ 0 & 0 & 1 & 0 \\ 0 & 0 & 0 & 1 \end{bmatrix}; \quad A_3 = \begin{bmatrix} c\theta_3 & 0 & s\theta_3 & 0 \\ s\theta_3 & 0 & -c\theta_3 & 0 \\ 0 & 1 & 0 & 0 \\ 0 & 0 & 0 & 1 \end{bmatrix}$$

$$A_4 = \begin{bmatrix} c\theta_4 & 0 & s\theta_4 & 0 \\ s\theta_4 & 0 & -c\theta_4 & 0 \\ 0 & 1 & 0 & c \\ 0 & 0 & 0 & 1 \end{bmatrix}; \quad A_5 = \begin{bmatrix} c\theta_5 & 0 & s\theta_5 & 0 \\ s\theta_5 & 0 & -c\theta_5 & 0 \\ 0 & 1 & 0 & 0 \\ 0 & 0 & 0 & 1 \end{bmatrix}; \quad A_6 = \begin{bmatrix} c\theta_6 & -s\theta_6 & 0 & 0 \\ s\theta_6 & c\theta_6 & 0 & 0 \\ 0 & 0 & 1 & h \\ 0 & 0 & 0 & 1 \end{bmatrix}$$

7.

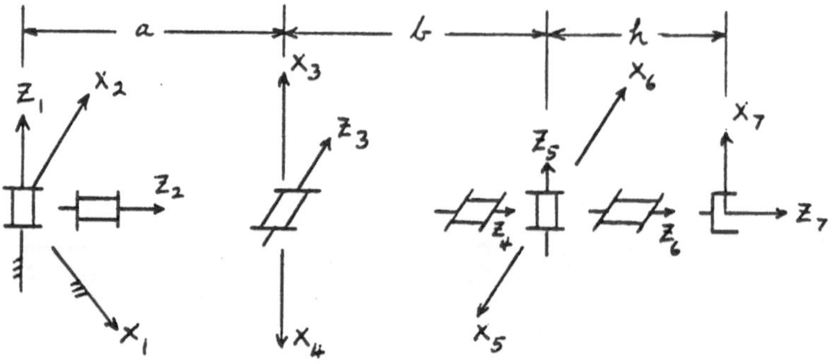

i	θ_i	s_i	α_i	a_i
1	$\theta_1(t)$	0	90°	0
2	$\theta_2(t)$	a	90°	0
3	$\theta_3(t)$	0	90°	0
4	$\theta_4(t)$	b	90°	0
5	$\theta_5(t)$	0	90°	0
6	$\theta_6(t)$	h	0	0

For the position shown,

$\theta_2 = 90°$, $\theta_3 = 180°$, $\theta_4 = -90°$,

$\theta_5 = 180°$, $\theta_6 = 90°$

$$A_1 = \begin{bmatrix} c\theta_1 & 0 & s\theta_1 & 0 \\ s\theta_1 & 0 & -c\theta_1 & 0 \\ 0 & 1 & 0 & 0 \\ 0 & 0 & 0 & 1 \end{bmatrix}; \quad A_2 = \begin{bmatrix} c\theta_2 & 0 & s\theta_2 & 0 \\ s\theta_2 & 0 & -c\theta_2 & 0 \\ 0 & 1 & 0 & a \\ 0 & 0 & 0 & 1 \end{bmatrix}; \quad A_3 = \begin{bmatrix} c\theta_3 & 0 & s\theta_3 & 0 \\ s\theta_3 & 0 & -c\theta_3 & 0 \\ 0 & 1 & 0 & 0 \\ 0 & 0 & 0 & 1 \end{bmatrix}$$

$$A_4 = \begin{bmatrix} c\theta_4 & 0 & s\theta_4 & 0 \\ s\theta_4 & 0 & -c\theta_4 & 0 \\ 0 & 1 & 0 & b \\ 0 & 0 & 0 & 1 \end{bmatrix}; \quad A_5 = \begin{bmatrix} c\theta_5 & 0 & s\theta_5 & 0 \\ s\theta_5 & 0 & -c\theta_5 & 0 \\ 0 & 1 & 0 & 0 \\ 0 & 0 & 0 & 1 \end{bmatrix}; \quad A_6 = \begin{bmatrix} c\theta_6 & -s\theta_6 & 0 & 0 \\ s\theta_6 & c\theta_6 & 0 & 0 \\ 0 & 0 & 1 & h \\ 0 & 0 & 0 & 1 \end{bmatrix}$$

8. $R = I + \mathbb{U}\sin\theta + \mathbb{U}^2(1-\cos\theta)$; $\underset{\sim}{d} = s\underset{\sim}{u} - (R-I)\underset{\sim}{Q}$

For $\theta = 30°$ & $\underset{\sim}{u} = \left(\frac{\sqrt{3}}{4}, \frac{3}{4}, \frac{1}{2}\right)$, $R = \begin{bmatrix} .89115 & -.2065 & .4040 \\ .2935 & .9414 & -.1663 \\ -.3459 & .2667 & .8995 \end{bmatrix}$

a) $\underset{\sim}{Q} = (0,0,0)$, $s = 0 \Rightarrow \underset{\sim}{d} \equiv \underset{\sim}{0}$

b) $\underset{\sim}{Q} = (1,1,0)$, $s = 0 \Rightarrow \underset{\sim}{d} = -(R-I)\underset{\sim}{Q} = \begin{bmatrix} .3153 \\ -.2349 \\ .0793 \end{bmatrix} \neq \underset{\sim}{0}$ (note!)

$$(D_{1\to 2})_a = \left[\begin{array}{ccc|c} & R & & \begin{matrix}0\\0\\0\end{matrix} \\ \hline 0 & 0 & 0 & 1 \end{array}\right] \quad \& \quad (D_{1\to 2})_b = \left[\begin{array}{ccc|c} & R & & \begin{matrix}.3153\\-.2349\\.0793\end{matrix} \\ \hline 0 & 0 & 0 & 1 \end{array}\right]$$

9. From equations (1.24) & (1.25), $\theta = 90°$, $\underset{\sim}{u} = (0,1,0)$

$s = \underset{\sim}{d}^t \underset{\sim}{u} = 1$

$s\underset{\sim}{u} - (R-I)\underset{\sim}{Q} = \underset{\sim}{d} \Rightarrow$
$Q_x - Q_z = 1$
$Q_x + Q_z = 3$
$1 = 1$ (drop)

Add $\underset{\sim}{u}^t \underset{\sim}{Q} = 0 \Rightarrow Q_y = 0$

Solution: $\underset{\sim}{Q} = (2, 0, 1)$

10.

$R_1 = R(\pi, \underset{\sim}{I}) = \begin{bmatrix} 1 & 0 & 0 \\ 0 & -1 & 0 \\ 0 & 0 & -1 \end{bmatrix}$ $\quad R_2\left(\pi, \frac{\underset{\sim}{I} + \underset{\sim}{J}}{2}\right) = \begin{bmatrix} 0 & 1 & 0 \\ 1 & 0 & 0 \\ 0 & 0 & -1 \end{bmatrix}$

Resultant $= R_2 R_1 = \begin{bmatrix} 0 & -1 & 0 \\ 1 & 0 & 0 \\ 0 & 0 & 1 \end{bmatrix}$; note that resultant is $R\left(\frac{\pi}{2}, \underset{\sim}{K}\right)$

11. (a) $R_{12} = I + W_{12}\sin\theta_{12} + W_{12}^2(1-\cos\theta_{12}) \approx I + W_{12}\theta_{12}$

$R_{23} = I + W_{23}\sin\theta_{23} + W_{23}^2(1-\cos\theta_{23}) \approx I + W_{23}\theta_{23}$

Combine: $R_{13} = R_{23}R_{12} \cong (I + W_{23}\theta_{23})(I + W_{12}\theta_{23})$

$\cong I + W_{12}\theta_{12} + W_{23}\theta_{23} +$ higher order terms

(b) For small θ_{12} & θ_{23}, the order of rotations can be changed without changing the resultant, i.e. $R_{13} = R_{23}R_{12} \equiv R_{12}R_{23}$.

But for large θ_{12} & θ_{13}, $R_{13} = R_{23}R_{12} \not\equiv R_{12}R_{23}$

12.

$$[D_{12}] = \begin{bmatrix} 1 & 0 & 0 & 0 \\ 0 & \cos\alpha & -\sin\alpha & 0 \\ 0 & \sin\alpha & \cos\alpha & 0 \\ 0 & 0 & 0 & 1 \end{bmatrix}$$

$$[D_{34}D_{23}] = \begin{bmatrix} 1 & 0 & 0 & a \\ 0 & 1 & 0 & 0 \\ 0 & 0 & 1 & s \\ 0 & 0 & 0 & 1 \end{bmatrix} \qquad [D_{14}] = \begin{bmatrix} 1 & 0 & 0 & a \\ 0 & \cos\alpha & -\sin\alpha & 0 \\ 0 & \sin\alpha & \cos\alpha & s \\ 0 & 0 & 0 & 1 \end{bmatrix}$$

$$[D_{45}] = \begin{bmatrix} \cos\theta & -\sin\theta & 0 & 0 \\ \sin\theta & \cos\theta & 0 & 0 \\ 0 & 0 & 1 & 0 \\ 0 & 0 & 0 & 1 \end{bmatrix} \qquad [D_{15}] = \begin{bmatrix} \cos\theta & -\sin\theta\cos\alpha & \sin\theta\sin\alpha & a\cos\theta \\ \sin\theta & \cos\theta\cos\alpha & -\cos\theta\sin\alpha & a\sin\theta \\ 0 & \sin\alpha & \cos\alpha & s \\ 0 & 0 & 0 & 1 \end{bmatrix}$$

$D_{15} = [D_{45}][D_{34}][D_{23}][D_{12}]$

The result is the standard DH matrix A (eqn 1.12 b), but it has been obtained from an "active" representation for displacements.

13. (a) $\mathcal{D}_e = \mathcal{D}_{1\to 2} = \mathcal{D}_2 \mathcal{D}_1^{-1}$ ("active" combination rule) ($\mathcal{D}_2$ is a displ. matrix)

$\Delta\mathcal{D} = \mathcal{D}_2 - \mathcal{D}_1$ (this difference is not a displacement matrix)

$[\nu] = [\mathcal{D}_e - I] = (\mathcal{D}_1 + \Delta\mathcal{D})\mathcal{D}_1^{-1} - I = [\Delta\mathcal{D}]\mathcal{D}_1^{-1}$

or $[\Delta\mathcal{D}] = [\nu]\mathcal{D}_1$

(b)

$$\mathcal{D}_e = \left[\begin{array}{c|c} R_e & \underset{\sim}{d}_e \\ \hline \underset{\sim}{0} & 1 \end{array}\right]$$

For small θ_e & S_e,

$$R_e = I + U_e \sin\theta_e + U_e^2(1-\cos\theta_e) \cong I + U_e\theta_e$$

$$\underset{\sim}{d}_e = S_e \underset{\sim}{u}_e - (R_e - I)\underset{\sim}{\rho}_e \cong S_e \underset{\sim}{u}_e - \theta_e U_e \underset{\sim}{\rho}_e$$

$$[\nu] = [\mathcal{D}_e - I] \simeq \left[\begin{array}{c|c} \theta_e U_e & S_e \underset{\sim}{u}_e - \theta_e U_e \underset{\sim}{\rho}_e \\ \hline \underset{\sim}{0} & 0 \end{array}\right]$$

14. $[\epsilon] = R_e - I = U_e \sin\theta_e + U_e^2(1-\cos\theta_e)$

$$[\epsilon^t\epsilon] = [-U_e \sin\theta_e + U_e^2(1-\cos\theta_e)][U_e \sin\theta_e + U_e^2(1-\cos\theta_e)]$$

$$= -U_e^2 \sin^2\theta_e + U_e^4(1-\cos\theta_e)^2 \quad ; \text{ note that } U_e^4 = -U_e^2$$

$$= -U_e^2\{\sin^2\theta_e + (1-\cos\theta_e)^2\}$$

$$= -2U_e^2(1-\cos\theta_e)$$

$$\text{Trace}[\epsilon^t\epsilon] = -2(1-\cos\theta_e).\text{Trace}[U_e^2]$$

$$= -2(1-\cos\theta_e).(-2u_x^2 - 2u_y^2 - 2u_z^2) \quad ; \text{ note } \underset{\sim}{u}^t\underset{\sim}{u} = 1$$

$$= 4(1-\cos\theta_e)$$

$$N(\epsilon) = \left(\sum_{i,j} \epsilon_{ij}^2\right)^{1/2} = \left(\text{Trace}[\epsilon^t\epsilon]\right)^{1/2} = 2.\sqrt{1-\cos\theta_e} \cong 1.414\,\theta_e$$

(for small θ_e)

15. $\Delta R = R_2 - R_1$ (difference ΔR is not a rotation matrix)

$\epsilon = R_e - I = R_2 R_1^t - I$; (R_e is a rotation matrix)

$\Delta R = R_2 - R_1 = (R_2 R_1^t - I) R_1 = [\epsilon] R_1$ (note typo in the book) (Compare with Prob. 13(a))

$[\Delta R]^t [\Delta R] = R_1^t [\epsilon^t \epsilon] R_1$

$\text{Trace}\{[\Delta R]^t [\Delta R]\} = \text{Trace}\{R_1^t [\epsilon^t \epsilon] R_1\}$

$= \text{Trace}[\epsilon^t \epsilon]$ (Trace of a matrix is not affected by a "similarity" operation)

$N(\Delta R) = \sqrt{\text{Trace}\{[\Delta R]^t [\Delta R]\}}$, $N(\epsilon) = \sqrt{\text{Trace}[\epsilon^t \epsilon]}$

$\therefore \; N(\Delta R) = N(\epsilon)$

$= 2 . \sqrt{1 - \cos \theta_e}$ (from Problem 14)

For fast computation, use $N(\Delta R) = \sqrt{\sum_{i,j} a_{ij}^2}$

16.(a) From eqn (1.31),

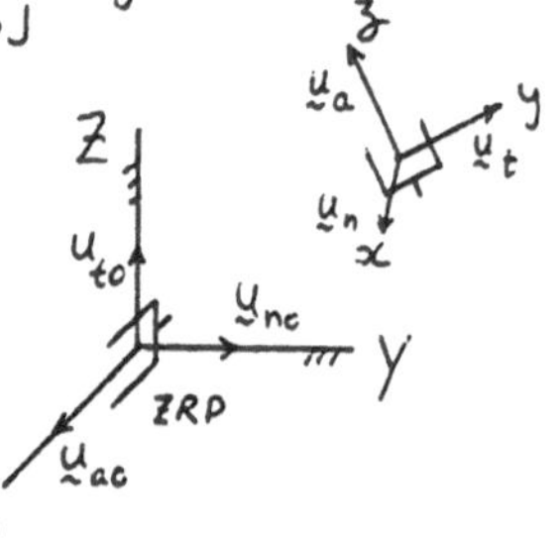

$$A_h = \begin{bmatrix} 2 & 2 & 2 & 3 \\ 3 & 3 & 0 & 3 \\ 2 & 0 & 0 & 0 \\ 1 & 1 & 1 & 1 \end{bmatrix} \begin{bmatrix} 3 & 3 & 0 & 3 \\ 0 & 2 & 2 & 2 \\ 1 & 1 & 1 & 0 \\ 1 & 1 & 1 & 1 \end{bmatrix}^{-1}$$

$$= \underset{(SAME)}{[\,"\,]} \cdot \begin{bmatrix} 0 & -1/2 & 0 & 1 \\ 1/3 & 1/2 & 1 & -2 \\ -1/3 & 0 & 0 & 1 \\ 0 & 0 & -1 & 1 \end{bmatrix} = \begin{bmatrix} 0 & 0 & -1 & 3 \\ 1 & 0 & 0 & 0 \\ 0 & -1 & 0 & 2 \\ 0 & 0 & 0 & 1 \end{bmatrix}$$

16. (b) From eq^n (1.33)

$$D_h = A_h \begin{bmatrix} \underset{\sim}{u}_{no} & \underset{\sim}{u}_{to} & \underset{\sim}{u}_{ao} & \underset{\sim}{O}\underset{\sim}{P}_o \\ 0 & 0 & 0 & 1 \end{bmatrix}^{-1}$$

$$= A_h \begin{bmatrix} 0 & 0 & 1 & 0 \\ 1 & 0 & 0 & 0 \\ 0 & 1 & 0 & 0 \\ 0 & 0 & 0 & 1 \end{bmatrix}^{-1}$$

$$= \begin{bmatrix} 0 & 0 & -1 & 3 \\ 1 & 0 & 0 & 0 \\ 0 & -1 & 0 & 2 \\ 0 & 0 & 0 & 1 \end{bmatrix} \begin{bmatrix} 0 & 1 & 0 & 0 \\ 0 & 0 & 1 & 0 \\ 1 & 0 & 0 & 0 \\ 0 & 0 & 0 & 1 \end{bmatrix} = \begin{bmatrix} -1 & 0 & 0 & 3 \\ 0 & 1 & 0 & 0 \\ 0 & 0 & -1 & 2 \\ 0 & 0 & 0 & 1 \end{bmatrix}$$

17. (a)

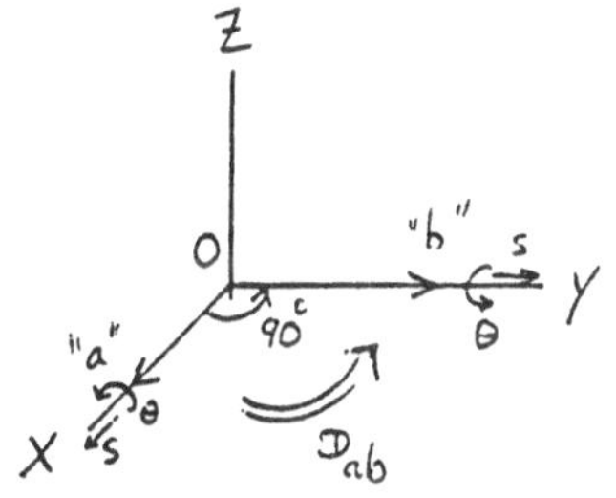

Let $D_a = D(\theta, s, \underset{\sim}{I}, \underset{\sim}{0}) = \begin{bmatrix} 1 & 0 & 0 & 0 \\ 0 & \cos\theta & -\sin\theta & 0 \\ 0 & \sin\theta & \cos\theta & 0 \\ 0 & 0 & 0 & 1 \end{bmatrix}$

$$D_{ab} = D(90^\circ, 0, \underset{\sim}{K}, \underset{\sim}{0}) = \begin{bmatrix} 0 & -1 & 0 & 0 \\ 1 & 0 & 0 & 0 \\ 0 & 0 & 1 & 0 \\ 0 & 0 & 0 & 1 \end{bmatrix}$$

$$D_{ab}^{-1} = \begin{bmatrix} 0 & 1 & 0 & 0 \\ -1 & 0 & 0 & 0 \\ 0 & 0 & 1 & 0 \\ 0 & 0 & 0 & 1 \end{bmatrix}$$

Then, $D_b = D_{ab}\, D_a\, D_{ab}^{-1} = \text{- - -} = \begin{bmatrix} \cos\theta & 0 & \sin\theta & 0 \\ 0 & 1 & 0 & 0 \\ -\sin\theta & 0 & \cos\theta & 0 \\ 0 & 0 & 0 & 1 \end{bmatrix} = D(\theta, s, \underset{\sim}{J}, \underset{\sim}{0})$

This is an exercise for using the principal of displacement similarity. Clearly, $D(\theta, s, \underset{\sim}{J}, \underset{\sim}{0})$ can be found from eq^{ns} (1.17) & (1.18).

17. (b)

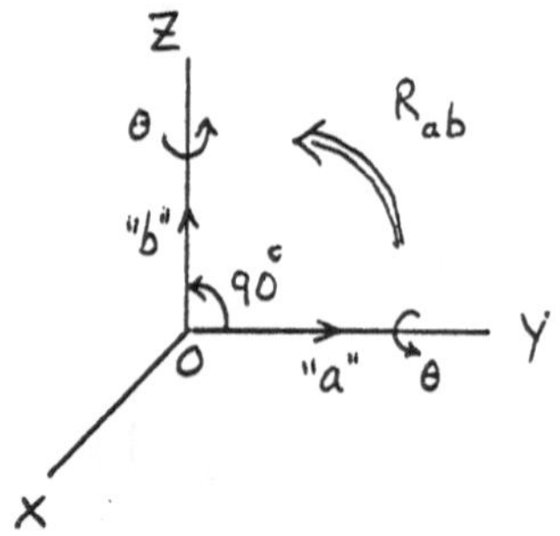

$$R_a = R(\theta, \underset{\sim}{J}) = \begin{bmatrix} \cos\theta & 0 & -\sin\theta \\ 0 & 1 & 0 \\ \sin\theta & 0 & \cos\theta \end{bmatrix}$$

$$R_{ab} = R(90^\circ, \underset{\sim}{I}) = \begin{bmatrix} 1 & 0 & 0 \\ 0 & 0 & -1 \\ 0 & 1 & 0 \end{bmatrix}$$

$$R_{ab}^t = \begin{bmatrix} 1 & 0 & 0 \\ 0 & 0 & 1 \\ 0 & -1 & 0 \end{bmatrix}$$

$$R_b = R_{ab} \cdot R_a \cdot R_{ab}^t = \text{---} = \begin{bmatrix} \cos\theta & -\sin\theta & 0 \\ \sin\theta & \cos\theta & 0 \\ 0 & 0 & 1 \end{bmatrix} = R(\theta, \underset{\sim}{K})$$

Again, this is an exercise on the use of rotational similarity. Clearly, $R(\theta, \underset{\sim}{K})$ can be found from eq^n(1.17) and it is given in eq^n(1.22).

CHAPTER 2

1. Cylindrical - 3 Roll Arm

i	Type	$\underset{\sim}{u}_{ic}$	$\underset{\sim}{Q}_{ic}$
1	R	0,0,1	0,0,0
2	P	0,0,1	0,0,0
3	P	1,0,0	0,0,0
4	R	1,0,0	0,0,0
5	R	0,0,1	0,0,0
6	R	1,0,0	0,0,0

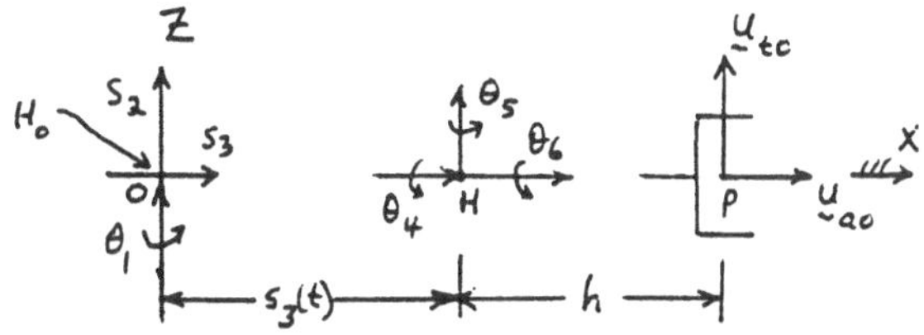

($S_3 = 0$ when H is at H_c)
Wrist-gripper not at ZRP for clarity

$\underset{\sim}{u}_{ao} = (1,0,0)$, $\underset{\sim}{u}_{to} = (0,0,1)$, $\underset{\sim}{P}_c = (h,0,0)$, $\underset{\sim}{H}_o = (0,0,0)$

Command: Move to $\underset{\sim}{P}$, $\underset{\sim}{u}_a$, $\underset{\sim}{u}_t$

$\underset{\sim}{H} = \underset{\sim}{P} - h\,\underset{\sim}{u}_a$; ($\underset{\sim}{H}$ is known)

$$D_1 = D(\theta_1, 0, \underset{\sim}{K}, \underset{\sim}{0}) = \begin{bmatrix} \cos\theta_1 & -\sin\theta_1 & 0 & 0 \\ \sin\theta_1 & \cos\theta_1 & 0 & 0 \\ 0 & 0 & 1 & 0 \\ 0 & 0 & 0 & 1 \end{bmatrix}$$

$$D_2 = D(0, S_2, \underset{\sim}{K}, \underset{\sim}{0}) = \begin{bmatrix} 1 & 0 & 0 & 0 \\ 0 & 1 & 0 & 0 \\ 0 & 0 & 1 & S_2 \\ 0 & 0 & 0 & 1 \end{bmatrix} \qquad D_3 = D(0, S_3, \underset{\sim}{I}, \underset{\sim}{0}) = \begin{bmatrix} 1 & 0 & 0 & S_3 \\ 0 & 1 & 0 & 0 \\ 0 & 0 & 1 & 0 \\ 0 & 0 & 0 & 1 \end{bmatrix}$$

$$\begin{bmatrix} \underset{\sim}{H} \\ 1 \end{bmatrix} = D_1 D_2 D_3 \begin{bmatrix} \underset{\sim}{H}_o \\ 1 \end{bmatrix} = \left[\begin{array}{ccc|c} \cos\theta_1 & -\sin\theta_1 & 0 & S_3\cos\theta_1 \\ \sin\theta_1 & \cos\theta_1 & 0 & S_3\sin\theta_1 \\ 0 & 0 & 1 & S_2 \\ \hline 0 & 0 & 0 & 1 \end{array}\right] \begin{bmatrix} 0 \\ 0 \\ 0 \\ 1 \end{bmatrix}$$

$$H_x = S_3 \cos\theta_1$$
$$H_y = S_3 \sin\theta_1$$
$$H_z = S_2$$

$\Rightarrow$ $S_2 = H_z$

$\theta_1 = \tan^{-1}(H_y/H_x)$; also $(\pi + \theta_1)$

$S_3 = H_x / \cos\theta_1$

1. (continued)

$$R_4 = R(\theta_4, \underset{\sim}{I}) = \begin{bmatrix} 1 & 0 & 0 \\ 0 & \cos\theta_4 & -\sin\theta_4 \\ 0 & \sin\theta_4 & \cos\theta_4 \end{bmatrix} ; \quad R_5 = R(\theta_5, \underset{\sim}{K}) = \begin{bmatrix} \cos\theta_5 & -\sin\theta_5 & 0 \\ \sin\theta_5 & \cos\theta_5 & 0 \\ 0 & 0 & 1 \end{bmatrix}$$

R_1, R_2, R_3 are principal 3x3 minors of D_1, D_2, D_3.

$$\underset{\sim}{v}_a = (R_1 R_2 R_3)^t \underset{\sim}{u}_a = R_4 R_5 \begin{bmatrix} 1 \\ 0 \\ 0 \end{bmatrix} \quad ; \quad (\underset{\sim}{v}_a \text{ is known})$$

$$v_{ax} = \cos\theta_5$$
$$v_{ay} = \cos\theta_4 \sin\theta_5$$
$$v_{az} = \sin\theta_4 \sin\theta_5$$

$$\theta_5 = \cos^{-1}(v_{ax}) \; ; \text{ also } (-\theta_5)$$

$$\theta_4 = 2\tan^{-1}\left(\frac{\sin\theta_4}{1+\cos\theta_4}\right) = 2\tan^{-1}\left(\frac{v_{az}/\sin\theta_5}{1 + v_{ay}/\sin\theta_5}\right) = 2\tan^{-1}\left(\frac{v_{az}}{\sin\theta_5 + v_{ay}}\right)$$

$$R_6 = R(\theta_6, \underset{\sim}{I}) = \begin{bmatrix} 1 & 0 & 0 \\ 0 & \cos\theta_6 & -\sin\theta_6 \\ 0 & \sin\theta_6 & \cos\theta_6 \end{bmatrix}$$

$$\underset{\sim}{w}_t = (R_4 R_5)^t (R_1 R_2 R_3)^t \underset{\sim}{u}_t = R_6 \begin{bmatrix} 0 \\ 0 \\ 1 \end{bmatrix} \quad ; \quad (\underset{\sim}{w}_t \text{ is known})$$

$$w_{ty} = -\sin\theta_6$$
$$w_{tz} = \cos\theta_6$$

$$\theta_6 = 2\tan^{-1}\left(\frac{\sin\theta_6}{1+\cos\theta_6}\right) = 2\tan^{-1}\left(\frac{-w_{ty}}{1+w_{tz}}\right)$$

2. Using Cramer's rule to solve for $\cos\theta_2$ and $\sin\theta_2$ (see page E1):

$$\cos\theta_2 = \frac{H_Z\, c.\sin\theta_3 - H_Q(b+c.\cos\theta_3)}{-(b^2+c^2+2bc\cos\theta_3)}, \quad \sin\theta_2 = \frac{H_Q\, c.\sin\theta_3 + H_Z(b+c.\cos\theta_3)}{-(b^2+c^2+2bc\cos\theta_3)}$$

substitute $b^2+c^2+2bc\cos\theta_3 = H_Z^2 + H_Q^2$ (for compactness)

$$\theta_2 = 2\tan^{-1}\left(\frac{\sin\theta_2}{1+\cos\theta_2}\right)$$

$$= 2\tan^{-1}\left(\frac{-cH_Q\sin\theta_3 - (b+c.\cos\theta_3)H_Z}{H_Q^2 + H_Z^2 + (b+c.\cos\theta_3)H_Q - cH_Z\sin\theta_3}\right)$$

3. On page E3:

substitute $\sin\theta_1 = \dfrac{2t}{1+t^2}$, $\cos\theta_1 = \dfrac{1-t^2}{1+t^2}$, $t = \tan\left(\dfrac{\theta_1}{2}\right)$

into $H_x\cos\theta_1 + H_y\sin\theta_1 = a + b\cos\theta_3$

$$\Rightarrow (H_x + a + b\cos\theta_3)t^2 - 2H_y t - (H_x - a - b\cos\theta_3) = 0$$

$$\Rightarrow \quad t = \frac{H_y \pm \sqrt{H_x^2 + H_y^2 - (a+b\cos\theta_3)^2}}{H_x + a + b\cos\theta_3}$$

Rewrite as $\theta_1 = 2\tan^{-1}\left(\dfrac{H_y + H_Q}{H_x + a + b\cos\theta_3}\right)$

where $H_Q = \pm\sqrt{H_x^2 + H_y^2 - (a+b\cos\theta_3)^2}$

4. On pages 66 & 67:

substitute $\cos\theta_1 = \frac{1-t^2}{1+t^2}$, $\sin\theta_1 = \frac{2t}{1+t^2}$, $t = \tan\left(\frac{\theta_1}{2}\right)$

into $H_y \cos\theta_1 - H_x \sin\theta_1 = a$

$\Rightarrow (a+H_y)t^2 + 2H_x t + (a - H_y) = 0$

$$\Rightarrow \quad t = \frac{-H_x \pm \sqrt{H_x^2 + H_y^2 - a^2}}{a + H_y}$$

Rewrite as $\theta_1 = 2\tan^{-1}\left(\frac{-H_x + H_R}{a + H_y}\right)$, where $H_R = \pm\sqrt{H_x^2 + H_y^2 - a^2}$

Use Cramer's rule to solve for $\cos\theta_2$ & $\sin\theta_2$:

$$\cos\theta_2 = \frac{-H_R(b - c\sin\theta_3) + cH_z\cos\theta_3}{-(b - c.\sin\theta_3)^2 - (c.\cos\theta_3)^2}$$

$$\sin\theta_2 = \frac{H_z(b - c.\sin\theta_3) + cH_R\cos\theta_3}{-(b - c.\sin\theta_3)^2 - (c.\cos\theta_3)^2}$$

$$\text{Denominator} = -(b - c.\sin\theta_3)^2 - (c.\cos\theta_3)^2$$

$$= 2bc\sin\theta_3 - b^2 - c^2$$

$$\theta_2 = 2\tan^{-1}\left(\frac{\sin\theta_2}{1 + \cos\theta_2}\right)$$

$$= 2\tan^{-1}\left(\frac{H_z(b - c.\sin\theta_3) + cH_R\cos\theta_3}{cH_z\cos\theta_3 - H_R(b - c.\sin\theta_3) + 2bc.\sin\theta_3 - b^2 - c^2}\right)$$

5.

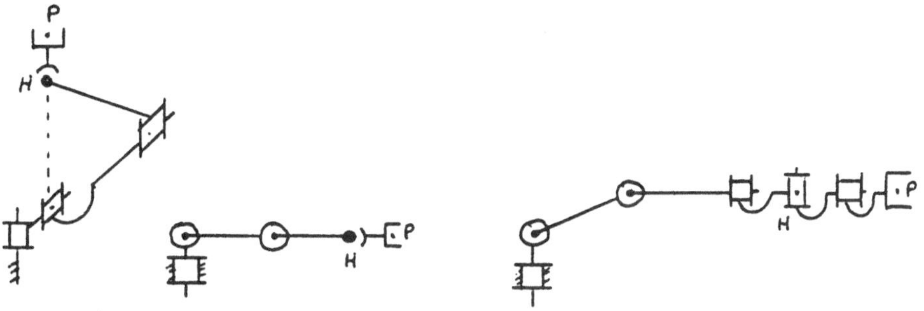

Wrist center above shoulder Elbow stretched Coplanar wrist axes

6. For PUMA, $a = 5$, $b = c = 20$, $h = 10$

$\underset{\sim}{P} = (20, 15, 0)$, $\underset{\sim}{u}_a = (0, 1, 0)$, $\underset{\sim}{u}_t = (0, 0, 1)$

$\underset{\sim}{H} = \underset{\sim}{P} - h\underset{\sim}{u}_a \Rightarrow H_x = 20, \quad H_y = 5, \quad H_z = 0$

$H_Q = \pm\sqrt{H_x^2 + H_y^2 - a^2} = \pm 20$

Then $\theta_1 = 0^\circ$ & -151.93°

$\theta_3 = 30^\circ$ & 150°

There are 4 combinations of these, but we need only 2 combinations with $\theta_1 = 0^\circ$. After long, but step-wise calculations, all angles are found as follows.

θ_1	θ_3	θ_2	θ_5	θ_4	θ_6
0°	30° →	-60° ↗	90° →	90° →	-60°
		↘	-90° →	-90° →	120°
0°	150° →	60° ↗	90° →	90° →	60°
		↘	-90° →	-90° →	-120°

7. (a) In section 2.6 (PUMA by ZRP),

$$\underset{\sim}{v}_a = \left\{ \begin{bmatrix} c\theta_1 & -s\theta_1 & 0 \\ s\theta_1 & c\theta_1 & 0 \\ 0 & 0 & 1 \end{bmatrix} \begin{bmatrix} c\theta_2 & 0 & +s\theta_2 \\ 0 & 1 & 0 \\ -s\theta_2 & 0 & c\theta_2 \end{bmatrix} \begin{bmatrix} c\theta_3 & 0 & +s\theta_3 \\ 0 & 1 & 0 \\ -s\theta_3 & 0 & c\theta_3 \end{bmatrix} \right\}^t \begin{bmatrix} u_{ax} \\ u_{ay} \\ u_{az} \end{bmatrix}$$

$$= \begin{bmatrix} \cos\theta_1 \cos(\theta_2+\theta_3) & \sin\theta_1 \cos(\theta_2+\theta_3) & -\sin(\theta_2+\theta_3) \\ -\sin\theta_1 & \cos\theta_1 & 0 \\ \cos\theta_1 \sin(\theta_2+\theta_3) & \sin\theta_1 \sin(\theta_2+\theta_3) & \cos(\theta_2+\theta_3) \end{bmatrix} \begin{bmatrix} u_{ax} \\ u_{ay} \\ u_{az} \end{bmatrix}$$

[MATRIX *]

$$= \begin{bmatrix} u_{ax} \cos\theta_1 \cos(\theta_2+\theta_3) + u_{ay} \sin\theta_1 \cos(\theta_2+\theta_3) - u_{az} \sin(\theta_2+\theta_3) \\ -u_{ax} \sin\theta_1 + u_{ay} \cos\theta_1 \\ u_{ax} \cos\theta_1 \sin(\theta_2+\theta_3) + u_{ay} \sin\theta_1 \sin(\theta_2+\theta_3) + u_{az} \cos(\theta_2+\theta_3) \end{bmatrix}$$

In section 2.7 (PUMA by Pieper-Roth method),

$$\underset{\sim}{v}'_a = \left\{ \begin{bmatrix} c\theta_1 & 0 & -s\theta_1 \\ s\theta_1 & 0 & c\theta_1 \\ 0 & -1 & 0 \end{bmatrix} \begin{bmatrix} c\theta_2 & -s\theta_2 & 0 \\ s\theta_2 & c\theta_2 & 0 \\ 0 & 0 & 1 \end{bmatrix} \begin{bmatrix} c\theta_3 & 0 & s\theta_3 \\ s\theta_3 & 0 & -c\theta_3 \\ 0 & 1 & 0 \end{bmatrix} \right\}^t \begin{bmatrix} u_{ax} \\ u_{ay} \\ u_{az} \end{bmatrix}$$

$$= \begin{bmatrix} \cos\theta_1 \cos(\theta_2+\theta_3) & \sin\theta_1 \cos(\theta_2+\theta_3) & -\sin(\theta_2+\theta_3) \\ -\sin\theta_1 & \cos\theta_1 & 0 \\ \cos\theta_1 \sin(\theta_2+\theta_3) & \sin\theta_1 \sin(\theta_2+\theta_3) & \cos(\theta_2+\theta_3) \end{bmatrix} \begin{bmatrix} u_{ax} \\ u_{ay} \\ u_{az} \end{bmatrix}$$

$$= \begin{bmatrix} u_{ax} \cos\theta_1 \cos(\theta_2+\theta_3) + u_{ay} \sin\theta_1 \cos(\theta_2+\theta_3) - u_{az} \sin(\theta_2+\theta_3) \\ -u_{ax} \sin\theta_1 + u_{ay} \cos\theta_1 \\ u_{ax} \cos\theta_1 \sin(\theta_2+\theta_3) + u_{ay} \sin\theta_1 \sin(\theta_2+\theta_3) + u_{az} \cos(\theta_2+\theta_3) \end{bmatrix}$$

$$\therefore \underset{\sim}{v}_a \equiv \underset{\sim}{v}'_a$$

7. (b) From section 2.6 (PUMA by ZRP),

$$\underset{\sim}{\omega}_t = \left\{ \begin{bmatrix} c\theta_4 & -s\theta_4 & 0 \\ s\theta_4 & c\theta_4 & 0 \\ 0 & 0 & 1 \end{bmatrix} \begin{bmatrix} c\theta_5 & 0 & -s\theta_5 \\ 0 & 1 & 0 \\ s\theta_5 & 0 & c\theta_5 \end{bmatrix} \right\}^t [\text{Matrix } *] \begin{bmatrix} u_{tx} \\ u_{ty} \\ u_{tz} \end{bmatrix}$$

$$= \begin{bmatrix} c\theta_4 c\theta_5 & s\theta_4 c\theta_5 & s\theta_5 \\ -s\theta_4 & c\theta_4 & 0 \\ -c\theta_4 s\theta_5 & -s\theta_4 s\theta_5 & c\theta_5 \end{bmatrix} [\text{Matrix } *] \begin{bmatrix} u_{tx} \\ u_{ty} \\ u_{tz} \end{bmatrix}$$

$$= \begin{bmatrix} \omega_{tx} \\ \omega_{ty} \\ \omega_{tz} \end{bmatrix}$$

From section 2.7,

$$\underset{\sim}{\omega}_t' = \left\{ \begin{bmatrix} c\theta_4 & 0 & s\theta_4 \\ s\theta_4 & 0 & -c\theta_4 \\ 0 & 1 & 0 \end{bmatrix} \begin{bmatrix} c\theta_5 & 0 & s\theta_5 \\ s\theta_5 & 0 & -c\theta_5 \\ 0 & 1 & 0 \end{bmatrix} \right\}^t [\text{Matrix } *] \begin{bmatrix} u_{tx} \\ u_{ty} \\ u_{tz} \end{bmatrix}$$

$$= \begin{bmatrix} c\theta_4 c\theta_5 & s\theta_4 c\theta_5 & s\theta_5 \\ s\theta_4 & -c\theta_4 & 0 \\ c\theta_4 s\theta_5 & s\theta_4 s\theta_5 & -c\theta_5 \end{bmatrix} [\text{Matrix } *] \begin{bmatrix} u_{tx} \\ u_{ty} \\ u_{tz} \end{bmatrix}$$

$$= \begin{bmatrix} \omega_{tx} \\ -\omega_{ty} \\ -\omega_{tz} \end{bmatrix}$$

$$\Rightarrow \omega_{tx}' = \omega_{tx}, \quad \omega_{ty}' = -\omega_{ty}, \quad \omega_{tz}' = -\omega_{tz}$$

8. From eqns (d), page 74,

$$x c_1 + y s_1 = (a_4 + l s_5) c_{234} + a_3 c_{23} + a_2 c_2$$

$$z = (a_4 + l s_5) s_{234} + a_3 s_{23} + a_2 s_2$$

Define $f = (a_4 + l s_5) c_{234} - x c_1 - y s_1$

$$g = (a_4 + l s_5) s_{234} - z$$

Then, $a_3 c_{23} = -f - a_2 c_2$

$$a_3 s_{23} = -g - a_2 s_2$$

Square and add: $a_3^2 = (f + a_2 c_2)^2 + (g + a_2 s_2)^2$

$$\Rightarrow (2a_2 g) s_2 + (2a_2 f) c_2 + (a_2^2 + f^2 + g^2 - a_3^2) = 0$$

or, $A \sin\theta_2 + B \cos\theta_2 + C = 0$

where $A = 2a_2 g$, $B = 2a_2 f$, $C = a_2^2 + f^2 + g^2 - a_3^2$

Substitute $\sin\theta_2 = \dfrac{2t}{1+t^2}$, $\cos\theta_2 = \dfrac{1-t^2}{1+t^2}$, $t = \tan\left(\dfrac{\theta_2}{2}\right)$

$$\Rightarrow (C-B)t^2 + 2At + (C+B) = 0$$

$$\Rightarrow t = \frac{-A \pm \sqrt{A^2+B^2-C^2}}{C-B}$$

or $\theta_2 = 2\tan^{-1}\left(\dfrac{-A \pm \sqrt{A^2+B^2-C^2}}{C-B}\right)$

8. (continued)

Rearrange 1st & 3rd eq^ns of set (g) on page 75:

$$\begin{bmatrix} c_{234} c_5 & -s_{234} \\ s_{234} c_5 & c_{234} \end{bmatrix} \begin{bmatrix} c_6 \\ s_6 \end{bmatrix} = \begin{bmatrix} u_{tx} c_1 + u_{ty} s_1 \\ u_{tz} \end{bmatrix}$$

Solve by Cramer's rule,

$$c_6 = \frac{c_{234} (u_{tx} c_1 + u_{ty} s_1) + s_{234} u_{tz}}{c_5}$$

$$s_6 = \frac{c_{234} c_5 u_{tz} - s_{234} c_5 (u_{tx} c_1 + u_{ty} s_1)}{c_5}$$

$$\theta_6 = 2 \tan^{-1} \left(\frac{s_6}{1 + c_6} \right)$$

$$= 2 \tan^{-1} \left[\frac{c_{234} c_5 u_{tz} - s_{234} c_5 (c_1 u_{tx} + s_1 u_{ty})}{c_5 + c_{234}(c_1 u_{tx} + s_1 u_{ty}) + s_{234} u_{tz}} \right]$$

9.

Center of the 5th joint lies on the axis of the 1st joint.

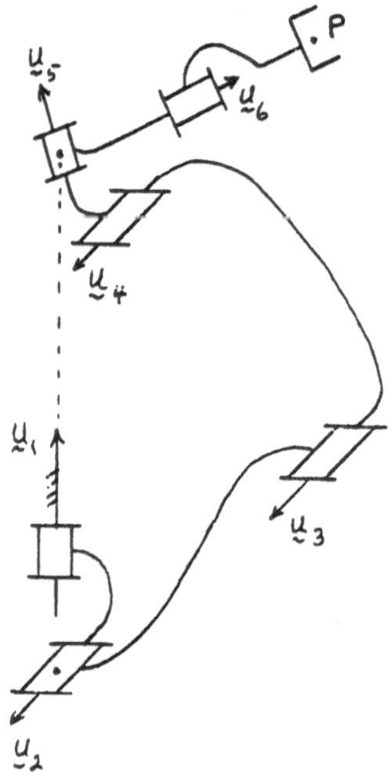

The other two singularity configurations are shown in Fig 2.6, page 72.

10. (a)

$$D_1 = \begin{bmatrix} c\theta_1 & -s\theta_1 & 0 & 0 \\ s\theta_1 & c\theta_1 & 0 & 0 \\ 0 & 0 & 1 & 0 \\ 0 & 0 & 0 & 1 \end{bmatrix}; \quad D_2 = \begin{bmatrix} c\theta_2 & 0 & s\theta_2 & 0 \\ 0 & 1 & 0 & 0 \\ -s\theta_2 & 0 & c\theta_2 & 0 \\ 0 & 0 & 0 & 1 \end{bmatrix}; \quad D_3 = \begin{bmatrix} c\theta_3 & -s\theta_3 & 0 & 12(1-c\theta_3) \\ s\theta_3 & c\theta_3 & 0 & -12\cdot s\theta_3 \\ 0 & 0 & 1 & 0 \\ 0 & 0 & 0 & 1 \end{bmatrix}$$

$$\begin{bmatrix} H_x \\ H_Y \\ H_Z \\ 1 \end{bmatrix} = D_1 D_2 D_3 \begin{bmatrix} 24'' \\ 0 \\ 0 \\ 1 \end{bmatrix}$$

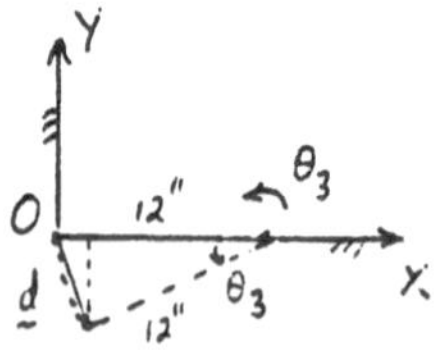

Expanding, we get the result given for H_x, H_Y, H_Z.

10. (b) $H_x^2 + H_y^2 + H_z^2 = (12)^2(1+\cos\theta_3)^2\cos^2\theta_2 + (12)^2\sin^2\theta_3 + (12)^2(1+\cos\theta_3)^2\sin^2\theta_2$

$$= (12)^2(1+\cos\theta_3)^2 + (12)^2\sin^2\theta_3$$

$$= 2\times(12)^2(1+\cos\theta_3)$$

$$\theta_3 = \cos^{-1}\left(\frac{H_x^2 + H_y^2 + H_z^2 - 288}{288}\right) \quad ; \text{ also } (-\theta_3)$$

Then, from the expression for H_z,

$$\theta_2 = \sin^{-1}\left[-\frac{H_z}{12(1+\cos\theta_3)}\right] \quad ; \text{ also } (\pi - \theta_2)$$

Rearranging the expressions for H_x and H_y:

$$\begin{bmatrix} 12(1+\cos\theta_3)\cos\theta_2 & -12\sin\theta_3 \\ 12\sin\theta_3 & 12(1+\cos\theta_3)\cos\theta_2 \end{bmatrix} \begin{bmatrix} \cos\theta_1 \\ \sin\theta_1 \end{bmatrix} = \begin{bmatrix} H_x \\ H_y \end{bmatrix}$$

$$\cos\theta_1 = \frac{H_x(1+\cos\theta_3)\cos\theta_2 + H_y\sin\theta_3}{12(1+\cos\theta_3)^2\cos^2\theta_2 + 12\sin^2\theta_2}$$

$$\sin\theta_1 = \frac{-H_x\sin\theta_3 + H_y(1+\cos\theta_3)\cos\theta_2}{12(1+\cos\theta_3)^2\cos^2\theta_2 + 12\sin^2\theta_2}$$

$$\theta_1 = 2\tan^{-1}\left(\frac{\sin\theta_1}{1+\cos\theta_1}\right)$$

$$= 2\tan^{-1}\left[\frac{-H_x\sin\theta_3 + H_y(1+\cos\theta_3)\cos\theta_2}{12(1+\cos\theta_3)^2\cos^2\theta_2 + 12\sin^2\theta_2 + H_x(1+\cos\theta_3)\cos\theta_2 + H_y\sin\theta_3}\right]$$

11. (a)

$$D_1 = D(\theta_1, 0, \underset{\sim}{K}, \underset{\sim}{O}) = \begin{bmatrix} \cos\theta_1 & -\sin\theta_1 & 0 & 0 \\ \sin\theta_1 & \cos\theta_1 & 0 & 0 \\ 0 & 0 & 1 & 0 \\ 0 & 0 & 0 & 1 \end{bmatrix}$$

$$D_2 = D(0, S_2, \underset{\sim}{I}, \underset{\sim}{O}) = \begin{bmatrix} 1 & 0 & 0 & S_2 \\ 0 & 1 & 0 & 0 \\ 0 & 0 & 1 & 0 \\ 0 & 0 & 0 & 1 \end{bmatrix}$$

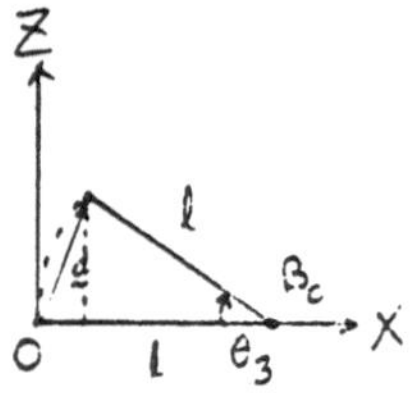

$$D_3 = D(\theta_3, 0, \underset{\sim}{J}, \underset{\sim}{B}_o) = \begin{bmatrix} \cos\theta_3 & 0 & \sin\theta_3 & l(1-\cos\theta_3) \\ 0 & 1 & 0 & 0 \\ -\sin\theta_3 & 0 & \cos\theta_3 & l\sin\theta_3 \\ 0 & 0 & 0 & 1 \end{bmatrix}$$

$$\begin{bmatrix} \underset{\sim}{H} \\ 1 \end{bmatrix} = D_1 D_2 D_3 \begin{bmatrix} \underset{\sim}{H}_c \\ 1 \end{bmatrix} = D_1 D_2 D_3 \begin{bmatrix} 2l \\ 0 \\ 0 \\ 1 \end{bmatrix} = \begin{bmatrix} \{l(1+\cos\theta_3) + S_2\}\cos\theta_1 \\ \{l(1+\cos\theta_3) + S_2\}\sin\theta_1 \\ -l\sin\theta_3 \\ 1 \end{bmatrix}$$

$$\therefore \quad H_x = \{l(1+\cos\theta_3) + S_2\}\cos\theta_1$$

$$H_y = \{l(1+\cos\theta_3) + S_2\}\sin\theta_1$$

$$H_z = -l\sin\theta_3$$

(b) There are two possible approaches.

Method I

$$\theta_3 = -\sin^{-1}(H_z/l) \; ; \; \text{also } (\pi - \theta_3)$$

$$\theta_1 = \tan^{-1}(H_y/H_x) \; ; \; \text{also } (\pi + \theta_1)$$

$$S_2 = (H_x/\cos\theta_1) - l(1+\cos\theta_3)$$

11. (b) (continued)

Method II

$$\theta_3 = -\sin^{-1}(H_z/\ell)\ ; \quad \text{also } (\pi - \theta_3)$$

Define $H_Q = \pm\sqrt{H_x^2 + H_y^2}$

Then, $H_Q = \ell(1 + \cos\theta_3) + S_2$

or, $S_2 = H_Q - \ell(1 + \cos\theta_3)$

From the expressions for H_x and H_y:

$$\theta_1 = 2\tan^{-1}\left(\frac{\sin\theta_1}{1+\cos\theta_1}\right) = 2\tan^{-1}\left(\frac{H_y}{H_x + \ell(1+\cos\theta_3) + S_2}\right)$$

(c) There are $2 \times 2 = 4$ distinct solutions. However, not all solutions can be realized physically because of mechanical interferences at the prismatic (#2) joint.

(d) when point H lies on the z-axis, joint angle θ_1 becomes undefined.

when joint angle $\theta_3 = \pm 90°$, the motion of point H in z-direction is not possible.

12. (a)

$$J'' = \left[\begin{array}{ccc|ccc} \underset{\sim}{u}_1 & \underset{\sim}{u}_2 & \underset{\sim}{0} & \underset{\sim}{u}_4 & \underset{\sim}{u}_5 & \underset{\sim}{u}_6 \\ \hline \underset{\sim}{u}_1 \times \underset{\sim}{Q}_1\underset{\sim}{H} & \underset{\sim}{u}_2 \times \underset{\sim}{Q}_2\underset{\sim}{H} & \underset{\sim}{u}_3 & \underset{\sim}{0} & \underset{\sim}{0} & \underset{\sim}{0} \end{array}\right]_{6\times 6} \equiv \left[\begin{array}{c|c} J_1 & J_3 \\ \hline J_2 & 0 \end{array}\right]$$

∴ Regional structure Jacobian $J_2 = \left[\underset{\sim}{u}_1 \times \underset{\sim}{Q}_1\underset{\sim}{H} \quad \underset{\sim}{u}_2 \times \underset{\sim}{Q}_2\underset{\sim}{H} \quad \underset{\sim}{u}_3\right]_{3\times 3}$

(b) Luckily, $\underset{\sim}{Q}_1 = \underset{\sim}{Q}_2 = \underset{\sim}{O}$. So, we need general expressions for $\underset{\sim}{u}_1, \underset{\sim}{u}_2, \underset{\sim}{u}_3$ and $\underset{\sim}{H}$.

$$\underset{\sim}{u}_1 = \underset{\sim}{K} = \begin{bmatrix} 0 \\ 0 \\ 1 \end{bmatrix}$$

$$\underset{\sim}{u}_2 = R_1 \underset{\sim}{u}_{20} = \begin{bmatrix} c\theta_1 & -s\theta_1 & 0 \\ s\theta_1 & c\theta_1 & 0 \\ 0 & 0 & 1 \end{bmatrix}\begin{bmatrix} 0 \\ 1 \\ 0 \end{bmatrix} = \begin{bmatrix} -s\theta_1 \\ c\theta_1 \\ 0 \end{bmatrix}$$

$$\underset{\sim}{u}_3 = R_1 R_2 \underset{\sim}{u}_{30} = \begin{bmatrix} c\theta_1 & -s\theta_1 & 0 \\ s\theta_1 & c\theta_1 & 0 \\ 0 & 0 & 1 \end{bmatrix}\begin{bmatrix} c\theta_2 & 0 & s\theta_2 \\ 0 & 1 & 0 \\ -s\theta_2 & 0 & c\theta_2 \end{bmatrix}\begin{bmatrix} 1 \\ 0 \\ 0 \end{bmatrix} = \begin{bmatrix} c\theta_1 c\theta_2 \\ s\theta_1 c\theta_2 \\ -s\theta_2 \end{bmatrix}$$

$$\begin{bmatrix} \underset{\sim}{H} \\ 1 \end{bmatrix} = D_1 D_2 D_3 \begin{bmatrix} \underset{\sim}{H}_0 \\ 1 \end{bmatrix}$$

$$= \begin{bmatrix} c\theta_1 & -s\theta_1 & 0 & 0 \\ s\theta_1 & c\theta_1 & 0 & 0 \\ 0 & 0 & 1 & 0 \\ 0 & 0 & 0 & 1 \end{bmatrix}\begin{bmatrix} c\theta_2 & 0 & s\theta_2 & 0 \\ 0 & 1 & 0 & 0 \\ -s\theta_2 & 0 & c\theta_2 & 0 \\ 0 & 0 & 0 & 1 \end{bmatrix}\begin{bmatrix} 1 & 0 & 0 & s_3 \\ 0 & 1 & 0 & 0 \\ 0 & 0 & 1 & 0 \\ 0 & 0 & 0 & 1 \end{bmatrix}\begin{bmatrix} 0 \\ 0 \\ 0 \\ 1 \end{bmatrix}$$

$$= \begin{bmatrix} s_3 \cdot c\theta_1 c\theta_2 \\ s_3 \cdot s\theta_1 c\theta_2 \\ -s_3 s\theta_2 \\ 1 \end{bmatrix}$$

12. (b) (continued)

Evaluate cross products; note that $\underset{\sim}{Q}_1 \underset{\sim}{H} = \underset{\sim}{H} - \overset{c}{\cancel{\underset{\sim}{c}_1}} = \underset{\sim}{H}$; also, $\underset{\sim}{Q}_2 \underset{\sim}{H} = \underset{\sim}{H}$

$$\underset{\sim}{u}_1 \times \underset{\sim}{Q}_1 \underset{\sim}{H} = \begin{vmatrix} \underset{\sim}{i} & \underset{\sim}{j} & \underset{\sim}{k} \\ 0 & 0 & 1 \\ H_x & H_y & H_z \end{vmatrix} = \begin{bmatrix} -S_3 . s\theta_1 c\theta_2 \\ S_3 . c\theta_1 c\theta_2 \\ 0 \end{bmatrix}$$

$$\underset{\sim}{u}_2 \times \underset{\sim}{Q}_2 \underset{\sim}{H} = \begin{vmatrix} \underset{\sim}{i} & \underset{\sim}{j} & \underset{\sim}{k} \\ -s\theta_1 & c\theta_1 & 0 \\ H_x & H_y & H_z \end{vmatrix} = \begin{bmatrix} -S_3 . c\theta_1 s\theta_2 \\ -S_3 . s\theta_1 s\theta_2 \\ -S_3 . c\theta_2 \end{bmatrix} \qquad [\nabla^t \underset{\sim}{H}]_{3\times 3} \swarrow$$

$$J_2 = \begin{bmatrix} -S_3 . s\theta_1 c\theta_2 & -S_3 c\theta_1 s\theta_2 & c\theta_1 c\theta_2 \\ S_3 . c\theta_1 c\theta_2 & -S_3 s\theta_1 s\theta_2 & s\theta_1 c\theta_2 \\ 0 & -S_3 c\theta_2 & -s\theta_2 \end{bmatrix}_{3\times 3} \equiv \begin{bmatrix} \frac{\partial H_x}{\partial \theta_1} & \frac{\partial H_x}{\partial \theta_2} & \frac{\partial H_x}{\partial S_3} \\ \frac{\partial H_y}{\partial \theta_1} & \frac{\partial H_y}{\partial \theta_2} & \frac{\partial H_y}{\partial S_3} \\ \frac{\partial H_z}{\partial \theta_1} & \frac{\partial H_z}{\partial \theta_2} & \frac{\partial H_z}{\partial S_3} \end{bmatrix}_{3\times 3}$$

(c)

$$\text{Det}\,|J_2| = -S_3 . s\theta_1 c\theta_2 \left[S_3 . s\theta_1 (s\theta_2)^2 + S_3 . s\theta_1 (c\theta_2)^2 \right]$$

$$- S_3 . c\theta_1 c\theta_2 \left[S_3 . c\theta_1 (s\theta_2)^2 + S_3 . c\theta_1 (c\theta_2)^2 \right]$$

$$= -S_3^2 (s\theta_1)^2 c\theta_2 - S_3^2 (c\theta_1)^2 c\theta_2$$

$$= -S_3^2 . c\theta_2$$

$\text{Det}\,|J_2| = 0$ when $S_3 = 0$ or $\theta_2 = \pm 90°$. Basically, when the arm is up/down, causing the wrist center H to be on the Z-axis, the point H can only move in the plane containing links 2 and 3. So, point H looses the ability to move perpendicular to the plane defined by links 2 and 3. Case $S_3 = 0$ is a special case of this situation where angle θ_2 becomes undefined.

13. (a) Using the active representation,

$$D_h' = D(\theta_2, 0, \underset{\sim}{u}_2, \underset{\sim}{s}_2) \cdot D(\theta_1, 0, \underset{\sim}{u}_{10}, \underset{\sim}{s}_{10})$$

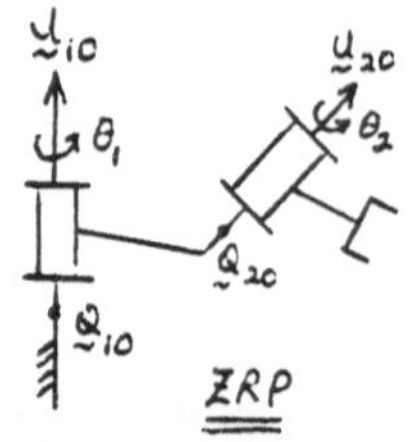

where $\underset{\sim}{u}_2 = R(\theta_1, \underset{\sim}{u}_{10}) \cdot \underset{\sim}{u}_{20}$ (shifted)

$$\begin{bmatrix} \underset{\sim}{s}_2 \\ 1 \end{bmatrix} = D(\theta_1, 0, \underset{\sim}{u}_{10}, \underset{\sim}{s}_{10}) \cdot \begin{bmatrix} \underset{\sim}{s}_{20} \\ 1 \end{bmatrix} \quad \text{(shifted)}$$

(b) $D_h'' = D(\theta_1, 0, \underset{\sim}{u}_{10}, \underset{\sim}{s}_{10}) \cdot D(\theta_2, 0, \underset{\sim}{u}_{20}, \underset{\sim}{s}_{20})$; note ZRP data $(\underset{\sim}{u}_{ic}, \underset{\sim}{s}_{ic})$

(c) If turn-slide wrt "a" is shifted to "b" due to D_{ab} (cause), then from the principle of displacement similarity,

$$D_b = D_{ab} D_a D_{ab}^{-1}$$

In part (a), axis $(\underset{\sim}{u}_{20}, \underset{\sim}{s}_{20})$ is shifted to $(\underset{\sim}{u}_2, \underset{\sim}{s}_2)$ due to the movement at the 1st joint : $D(\theta_1, 0, \underset{\sim}{u}_{10}, \underset{\sim}{s}_{10}) \equiv D_{ab}$

$$\Rightarrow D(\theta_2, 0, \underset{\sim}{u}_2, \underset{\sim}{s}_2) = D(\theta_1, 0, \underset{\sim}{u}_{10}, \underset{\sim}{s}_{10}) \cdot D(\theta_2, 0, \underset{\sim}{u}_{20}, \underset{\sim}{s}_{20}) \cdot D^{-1}(\theta_1, 0, \underset{\sim}{u}_{10}, \underset{\sim}{s}_{10})$$

Substituting into D_h':

$$D_h' = \cdots = D(\theta_1, 0, \underset{\sim}{u}_{10}, \underset{\sim}{s}_{10}) \cdot D(\theta_2, 0, \underset{\sim}{u}_{20}, \underset{\sim}{s}_{20}) = D_h''$$

For n-jointed serial robots, the result of part (b) can be generalized as (see eq^n (2.5)):

$$D_{hand} = \prod_{i=1}^{n} D(\theta_i, s_i, \underset{\sim}{u}_{ic}, \underset{\sim}{s}_{ic}) \quad \text{where } D_{hand} = \left[D_{ZRP \to CP} \right]_{hand}$$

and $\prod_{i=1}^{n}$ is the product symbol.

14.
$$\begin{bmatrix}\theta_4\\ \theta_5\\ \theta_6\end{bmatrix} = \begin{bmatrix}1 & -1/t & -1\\ 1 & 1/t & -1\\ 1 & 0 & 0\end{bmatrix}^{-1}\begin{bmatrix}\phi_1\\ \phi_2\\ \phi_3\end{bmatrix} = (t/2)\begin{bmatrix}0 & -1 & -1/t\\ 0 & 1 & -1/t\\ 2/t & 0 & 2/t\end{bmatrix}^{t}\begin{bmatrix}\phi_1\\ \phi_2\\ \phi_3\end{bmatrix}$$

$$= \begin{bmatrix}0 & 0 & 1\\ -t/2 & t/2 & 0\\ -1/2 & -1/2 & 1\end{bmatrix}\begin{bmatrix}\phi_1\\ \phi_2\\ \phi_3\end{bmatrix}$$

15. Using $\beta = 20^\circ$ in Table 2.9:

Identical gears G_1 & $G_3 \Rightarrow \alpha_1 = 90^\circ - \beta - \alpha_1$

or, $\alpha_1 = \frac{1}{2}(90^\circ - \beta) = 35^\circ$

Identical gears G_4 & $G_6 \Rightarrow \alpha_3 = 90^\circ - \beta - \alpha_3$

or, $\alpha_3 = \frac{1}{2}(90^\circ - \beta) = 35^\circ$

Half pitch cone angle for gear $G_5 = 90^\circ \equiv 90^\circ + \beta - \alpha_2$

or, $\alpha_2 = \beta = 20^\circ$

Gear #	G_1	G_2	G_3	G_4	G_5	G_6
Half pitch cone angle	35°	20°	35°	35°	90°	35°

16. Following the solution for problem 15, with $\beta = 30^\circ$ & half cone angle for gear $G_5 = 90^\circ$,

$\alpha_1 = \alpha_3 = 30^\circ$, $\alpha_2 = 30^\circ$

Gear #	G_1	G_2	G_3	G_4	G_5	G_6
Half pitch cone angle	30°	30°	30°	30°	90°	30°

16. (continued)

$$\begin{bmatrix} \phi_1 \\ \phi_2 \\ \phi_3 \end{bmatrix} = \begin{bmatrix} 1 & -\dfrac{\cos(\alpha_1+\beta)}{\sin\alpha_1} & \dfrac{\sin\alpha_3 \cos(\alpha_1+\beta)}{\cos(\alpha_3+\beta)\sin\alpha_1} \\ 1 & \dfrac{\cos(\alpha_2-\beta)}{\sin\alpha_2} & 0 \\ 1 & 0 & 0 \end{bmatrix} \begin{bmatrix} \theta_4 \\ \theta_5 \\ \theta_6 \end{bmatrix}$$

$$= \begin{bmatrix} 1 & -1 & 1 \\ 1 & 2 & 0 \\ 1 & 0 & 0 \end{bmatrix} \begin{bmatrix} \theta_4 \\ \theta_5 \\ \theta_6 \end{bmatrix}$$

Inverse relation:

$$\begin{bmatrix} \theta_4 \\ \theta_5 \\ \theta_6 \end{bmatrix} = \begin{bmatrix} 0 & 0 & 1 \\ 0 & 1/2 & -1/2 \\ 1 & 1/2 & -3/2 \end{bmatrix} \begin{bmatrix} \phi_1 \\ \phi_2 \\ \phi_3 \end{bmatrix}$$

17.

$$\underset{\sim}{v}_{o'} = \underset{\sim}{v}_p - \underset{\sim}{\omega} \times \underset{\sim}{O'P} \qquad (\text{Eq. } 2.22)$$

$$= \begin{bmatrix} 10 \\ -3 \\ 5 \end{bmatrix} - \begin{bmatrix} 17 \\ -7 \\ 1 \end{bmatrix} = \begin{bmatrix} -7 \\ 4 \\ 4 \end{bmatrix}$$

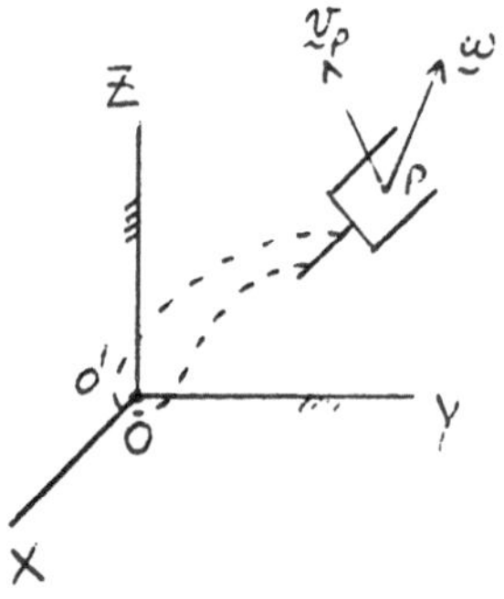

$$[V] = \left[\begin{array}{c|c} \Omega & \underset{\sim}{v}_{o'} \\ \hline \underset{\sim}{0} & 0 \end{array}\right] = \begin{bmatrix} 0 & -\omega_z & \omega_y & v_{o'x} \\ \omega_z & 0 & -\omega_x & v_{o'y} \\ -\omega_y & \omega_x & 0 & v_{o'z} \\ 0 & 0 & 0 & 0 \end{bmatrix} = \begin{bmatrix} 0 & -1 & 5 & -7 \\ 1 & 0 & -2 & 4 \\ -5 & 2 & 0 & 4 \\ 0 & 0 & 0 & 0 \end{bmatrix}$$

18. (a) $\omega = 10$ rad/s , $v_{ISA} = 2$ in/s , $\underset{\sim}{\rho}_{ISA} = (2, 3, 0)^t$

$$\underset{\sim}{u}_{ISA} = \begin{bmatrix} .07018 \\ .40351 \\ .91228 \end{bmatrix}_{3\times1} \quad ; \quad U_{ISA} = \begin{bmatrix} 0 & -.91228 & .40351 \\ .91228 & 0 & -.07018 \\ -.40351 & .07018 & 0 \end{bmatrix}_{3\times3}$$

$$[\Omega] = \omega U_{ISA} = \begin{bmatrix} 0 & -9.1228 & 4.0351 \\ 9.1228 & 0 & -.7018 \\ -4.0351 & .7018 & 0 \end{bmatrix}$$

$$[V] = \left[\begin{array}{c|c} \Omega & v_{ISA}\underset{\sim}{u}_{ISA} - \omega U_{ISA}\underset{\sim}{\rho}_{ISA} \\ \hline \underset{\sim}{0} & 0 \end{array}\right] \qquad (\text{eq}^n\ 2.29)$$

$$= \begin{bmatrix} 0 & -9.1228 & 4.0351 & 27.50876 \\ 9.1228 & 0 & -.7018 & -17.43858 \\ -4.0351 & .7018 & 0 & 7.78936 \\ 0 & 0 & 0 & 0 \end{bmatrix}$$

(b) $$\begin{bmatrix} \underset{\sim}{v}_p \\ 0 \end{bmatrix} = [V]\begin{bmatrix} \underset{\sim}{p} \\ 1 \end{bmatrix} \qquad (\text{eq}^n\ 2.30)$$

For point P: (1, 1, 4), $\underset{\sim}{v}_p = (34.52636, -11.12298, 4.45666)^t$

For point P: (2.4, 5.3, 5.2), $\underset{\sim}{v}_p = (.14044, .80678, 1.82466)^t$

This point is on ISA, so this $\underset{\sim}{v}_p \equiv v_{ISA}\,\underset{\sim}{u}_{ISA}$ (checks)

For point P: (0, 0, 0), $\underset{\sim}{v}_p = (27.50876, -17.43858, 7.78936)^t$

This $\underset{\sim}{v}_p$ is same as $\underset{\sim}{v}_{0'}$ (0' is body point coincident with the base origin)

19. (a)

$$D_i = D(\theta_i, s_i, \mathbf{u}_{ic}, \mathbf{Q}_{ic}) = \left[\begin{array}{c|c} R_i & \mathbf{d}_i \\ \hline \mathbf{0} & 1 \end{array}\right]$$

$$\dot{D}_i = \left[\begin{array}{c|c} \dot{R}_i & \dot{\mathbf{d}}_i \\ \hline \mathbf{0} & 0 \end{array}\right]$$

$$D_i^{-1} = \left[\begin{array}{c|c} R_i^t & -R_i^t \mathbf{d}_i \\ \hline \mathbf{0} & 1 \end{array}\right]$$

$$\dot{D}_i D_i^{-1} = \left[\begin{array}{c|c} \dot{R}_i R_i^t & -\dot{R}_i R_i^t \mathbf{d}_i + \dot{\mathbf{d}}_i \\ \hline \mathbf{0} & 0 \end{array}\right]$$

$$R_i = I + U_{io} \sin\theta_i + U_{io}^2 (1 - \cos\theta_i)$$

$$\dot{R}_i = \dot{\theta}_i U_{ic} \cos\theta_i + \dot{\theta}_i U_{io}^2 \sin\theta_i \qquad (U_{ic} \text{ is ZRP data; it is not differentiated})$$

$$\dot{R}_i R_i^t = \dot{\theta}_i \left(U_{ic} \cos\theta_i + U_{ic}^2 \sin\theta_i \right) \left(I - U_{io} \sin\theta_i + U_{io}^2 (1 - \cos\theta_i) \right)$$

$$= \dot{\theta}_i \Big(U_{ic} \cos\theta - U_{ic}^2 \sin\theta_i \cos\theta_i + U_{ic}^3 \cos\theta_i (1 - \cos\theta_i)$$

$$+ U_{ic}^2 \sin\theta_i - U_{ic}^3 \sin^2\theta_i + U_{ic}^4 \sin\theta_i (1 - \cos\theta_i) \Big)$$

using $U^3 = -U$, $U^4 = -U^2$,

$$\dot{R}_i R_i^t = \cdots = \dot{\theta}_i U_{ic}$$

Next, $\mathbf{d}_i = s_i \mathbf{u}_{io} - (R_i - I)\mathbf{Q}_{io}$ & $\dot{\mathbf{d}}_i = \dot{s}_i \mathbf{u}_{ic} - \dot{R}\mathbf{Q}_{ic}$ ($\mathbf{u}_{ic}$ & $\mathbf{Q}_{ic}$ = ZRP data)

$$-\dot{R}_i R_i^t \mathbf{d}_i + \dot{\mathbf{d}}_i = -\dot{\theta}_i U_{io} \{ s_i \mathbf{u}_{ic} - (R_i - I)\mathbf{Q}_{ic} \} + \dot{s}_i \mathbf{u}_{ic} - \dot{\theta}(U_{ic} \cos\theta_i + U_{ic}^2 \sin\theta_i) \mathbf{Q}_{ic}$$

19.(a) (continued)

$$-\dot{R}_i R_i^t \underset{\sim}{d}_i + \dot{\underset{\sim}{d}}_i = -\dot{\theta}_i s_i \underbrace{\cancel{U_{ic}\underset{\sim}{u}_{ic}}}_{0} + \dot{\theta}_i\left(\cancel{U_{ic}^2 \sin\theta_i} + U_{io}^3(1-\cancel{c}\theta_i)\right)\underset{\sim}{Q}_{ic}$$

$$+\, \dot{s}_i \underset{\sim}{u}_{ic} - \dot{\theta}_i\left(U_{ic}\cancel{\cos}\theta_i + \cancel{U_{io}^2 \sin\theta_i}\right)\underset{\sim}{Q}_{ic}$$

Note that $U^3 = -U$ and $U\underset{\sim}{u} = 0$ (i.e. $\underset{\sim}{u} \times \underset{\sim}{u} = 0$).

Finally, $-\dot{R}_i R_i^t \underset{\sim}{d}_i + \dot{\underset{\sim}{d}}_i = \dot{s}_i \underset{\sim}{u}_{ic} - \dot{\theta}_i U_{ic} \underset{\sim}{Q}_{ic}$

and $$[V] = \dot{D}_i D_i^{-1} = \left[\begin{array}{c|c} \dot{\theta}_i U_{ic} & \dot{s}_i \underset{\sim}{u}_{ic} - \dot{\theta}_i U_{ic}\underset{\sim}{Q}_{ic} \\ \hline \underset{\sim}{0} & 0 \end{array}\right]$$

Also, directly from eqn (2.29), with $\underset{\sim}{u}_{ISA} \equiv \underset{\sim}{u}_{ic}$ and $\underset{\sim}{Q}_{ISA} \equiv \underset{\sim}{Q}_{ic}$,

$$[V] = [V(\dot{\theta}_i, \dot{s}_i, \underset{\sim}{u}_{io}, \underset{\sim}{Q}_{io})] = \left[\begin{array}{c|c} \dot{\theta}_i U_{ic} & \dot{s}_i \underset{\sim}{u}_{io} - \dot{\theta}_i U_{ic}\underset{\sim}{Q}_{ic} \\ \hline \underset{\sim}{0} & 0 \end{array}\right]$$

(b)

$$B_i = [V(1, 0, \underset{\sim}{u}_{io}, \underset{\sim}{Q}_{io})] = \left[\begin{array}{c|c} U_{ic} & -U_{ic}\underset{\sim}{Q}_{ic} \\ \hline \underset{\sim}{0} & 0 \end{array}\right]$$

This is a constant matrix associated with the ith revolute joint.

$$B_j = [V(0, 1, \underset{\sim}{u}_{jo}, \underset{\sim}{Q}_{jo})] = \left[\begin{array}{c|c} 0_{3\times 3} & \underset{\sim}{u}_{jc} \\ \hline \underset{\sim}{0} & 0 \end{array}\right]$$

This is a constant matrix associated with the jth prismatic joint.

20. $$J_2 = \begin{bmatrix} \underset{\sim}{u}_1 \times \underset{\sim}{Q_1 H} & \underset{\sim}{u}_2 \times \underset{\sim}{Q_2 H} & \underset{\sim}{u}_3 \times \underset{\sim}{Q_3 H} \end{bmatrix}_{3\times 3}$$

$$D_1 = \begin{bmatrix} c\theta_1 & -s\theta_1 & 0 & 0 \\ s\theta_1 & c\theta_1 & 0 & 0 \\ 0 & 0 & 1 & 0 \\ 0 & 0 & 0 & 1 \end{bmatrix} ; \quad D_2 = \begin{bmatrix} c\theta_2 & 0 & s\theta_2 & -a(1-c\theta_2) \\ 0 & 1 & 0 & 0 \\ -s\theta_2 & 0 & c\theta_2 & -a\,s\theta_2 \\ 0 & 0 & 0 & 1 \end{bmatrix}$$

$$D_3 = \begin{bmatrix} c\theta_3 & 0 & s\theta_3 & (b-a)(1-c\theta_3) \\ 0 & 1 & 0 & 0 \\ -s\theta_3 & 0 & c\theta_3 & (b-a)s\theta_3 \\ 0 & 0 & 0 & 1 \end{bmatrix}$$

; R_1, R_2, R_3 are principal 3×3 minors of D_1, D_2, D_3.

$$\underset{\sim}{u}_1 \equiv \underset{\sim}{u}_{10} = \begin{bmatrix} 0 \\ 0 \\ 1 \end{bmatrix}$$

$$\underset{\sim}{u}_2 = R_1 \underset{\sim}{u}_{20} = R_1 \begin{bmatrix} 0 \\ 1 \\ 0 \end{bmatrix} = \begin{bmatrix} -s\theta_1 \\ c\theta_1 \\ 0 \end{bmatrix}$$

$$\underset{\sim}{u}_3 = R_1 R_2 \underset{\sim}{u}_{30} = R_1 R_2 \begin{bmatrix} 0 \\ 1 \\ 0 \end{bmatrix} = \begin{bmatrix} -s\theta_1 \\ c\theta_1 \\ 0 \end{bmatrix}$$

$$\underset{\sim}{Q}_1 \equiv \underset{\sim}{Q}_{10} = \begin{bmatrix} 0 \\ 0 \\ 0 \end{bmatrix}$$

$$\begin{bmatrix} \underset{\sim}{Q}_2 \\ 1 \end{bmatrix} = D_1 \begin{bmatrix} \underset{\sim}{Q}_{20} \\ 1 \end{bmatrix} = D_1 \begin{bmatrix} -a \\ 0 \\ 0 \\ 1 \end{bmatrix} = \begin{bmatrix} -a\,c\theta_1 \\ -a\,s\theta_1 \\ 0 \\ 1 \end{bmatrix}$$

20. (continued)

$$\begin{bmatrix} \underset{\sim}{Q}_3 \\ 1 \end{bmatrix} = D_1 D_2 \begin{bmatrix} \underset{\sim}{Q}_{30} \\ 1 \end{bmatrix} = D_1 D_2 \begin{bmatrix} b-a \\ 0 \\ 0 \\ 1 \end{bmatrix} = \begin{bmatrix} b\,c\theta_1 c\theta_2 - a\,c\theta_1 \\ b\,s\theta_1 c\theta_2 - a\,s\theta_1 \\ -b\,s\theta_2 \\ 1 \end{bmatrix}$$

$$\begin{bmatrix} \underset{\sim}{H} \\ 1 \end{bmatrix} = D_1 D_2 D_3 \begin{bmatrix} \underset{\sim}{H}_0 \\ 1 \end{bmatrix} = D_1 D_2 D_3 \begin{bmatrix} b+c-a \\ 0 \\ 0 \\ 1 \end{bmatrix} = \begin{bmatrix} c\theta_1 \{c.\cos(\theta_2+\theta_3) + b\,c\theta_2 - a\} \\ s\theta_1 \{c.\cos(\theta_2+\theta_3) + b\,c\theta_2 - a\} \\ -c.\sin(\theta_2+\theta_3) - b\,s\theta_2 \\ 1 \end{bmatrix}$$

$\underset{\sim}{Q}_1\underset{\sim}{H} = \underset{\sim}{H} - \underset{\sim}{Q}_1 = \underset{\sim}{H}$, $\underset{\sim}{Q}_2\underset{\sim}{H} = \underset{\sim}{H} - \underset{\sim}{Q}_2$, $\underset{\sim}{Q}_3\underset{\sim}{H} = \underset{\sim}{H} - \underset{\sim}{Q}_3$

Evaluate $\underset{\sim}{u}_1 \times \underset{\sim}{Q}_1\underset{\sim}{H}$, $\underset{\sim}{u}_2 \times \underset{\sim}{Q}_2\underset{\sim}{H}$, $\underset{\sim}{u}_3 \times \underset{\sim}{Q}_3\underset{\sim}{H}$ (this is lot of work!)

$$J_2 = \left[\begin{array}{c|c|c} -\sin\theta_1\{c.\cos(\theta_2+\theta_3)+b\cos\theta_2-a\} & -\cos\theta_1\{c.\sin(\theta_2+\theta_3)+b\sin\theta_2\} & -c.\cos\theta_1\sin(\theta_2+\theta_3) \\ \cos\theta_1\{c.\cos(\theta_2+\theta_3)+b\cos\theta_2-a\} & -\sin\theta_1\{c.\sin(\theta_2+\theta_3)+b\sin\theta_2\} & -c.\sin\theta_1\sin(\theta_2+\theta_3) \\ 0 & -\{c.\cos(\theta_2+\theta_3)+b\cos\theta_2\} & -c.\cos(\theta_2+\theta_3) \end{array}\right]$$

$\det|J_2| = bc\sin\theta_3\{c.\cos(\theta_2+\theta_3) + b\cos\theta_2 - a\}$ (again lot of work)

Simpler way to find J_2:

$$J_2 = \nabla^t \underset{\sim}{H} = \begin{bmatrix} \frac{\partial H_x}{\partial\theta_1} & \frac{\partial H_x}{\partial\theta_2} & \frac{\partial H_x}{\partial\theta_3} \\ \frac{\partial H_y}{\partial\theta_1} & \frac{\partial H_y}{\partial\theta_2} & \frac{\partial H_y}{\partial\theta_3} \\ \frac{\partial H_z}{\partial\theta_1} & \frac{\partial H_z}{\partial\theta_2} & \frac{\partial H_z}{\partial\theta_3} \end{bmatrix}$$

21. $$J_2 = \left[\underset{\sim}{u}_1 \times \underset{\sim}{Q}_1 \underset{\sim}{H} \qquad \underset{\sim}{u}_2 \times \underset{\sim}{Q}_2 \underset{\sim}{H} \qquad \underset{\sim}{u}_3 \times \underset{\sim}{Q}_3 \underset{\sim}{H}\right]_{3\times 3}$$

$$D_1 = \begin{bmatrix} c_1 & -s_1 & 0 & 0 \\ s_1 & c_1 & 0 & 0 \\ 0 & 0 & 1 & 0 \\ 0 & 0 & 0 & 1 \end{bmatrix}; \quad D_2 = \begin{bmatrix} 1 & 0 & 0 & 0 \\ 0 & c_2 & -s_2 & 0 \\ 0 & s_2 & c_2 & 0 \\ 0 & 0 & 0 & 1 \end{bmatrix}; \quad D_3 = \begin{bmatrix} c_3 & 0 & s_3 & a(1-c_3) \\ 0 & 1 & 0 & 0 \\ -s_3 & 0 & c_3 & a s_3 \\ 0 & 0 & 0 & 1 \end{bmatrix}$$

$$\begin{bmatrix} \underset{\sim}{H} \\ 1 \end{bmatrix} = D_1 D_2 D_3 \begin{bmatrix} \underset{\sim}{H}_0 \\ 1 \end{bmatrix} = D_1 D_2 D_3 \begin{bmatrix} a+b \\ 0 \\ 0 \\ 1 \end{bmatrix} = \begin{bmatrix} (a+bc_3)c_1 - b s_1 s_2 s_3 \\ (a+bc_3)s_1 + b c_1 s_2 s_3 \\ -b c_2 s_3 \\ 1 \end{bmatrix}$$

$$\underset{\sim}{u}_1 = \underset{\sim}{u}_{10} = \begin{bmatrix} 0 \\ 0 \\ 1 \end{bmatrix}, \quad \underset{\sim}{Q}_1 = \underset{\sim}{Q}_{10} = \begin{bmatrix} 0 \\ 0 \\ 0 \\ 1 \end{bmatrix}$$

Note – R_1, R_2, R_3 are principal 3×3 minors of $D_1 D_2 D_3$

$$\underset{\sim}{u}_1 \times \underset{\sim}{Q}_1 \underset{\sim}{H} = \underset{\sim}{u}_1 \times \underset{\sim}{H} = \begin{bmatrix} -H_y \\ H_x \\ 0 \end{bmatrix}$$

$$\underset{\sim}{u}_2 = R_1 \underset{\sim}{u}_{20} = R_1 \begin{bmatrix} 1 \\ 0 \\ 0 \end{bmatrix} = \begin{bmatrix} c_1 \\ s_1 \\ 0 \end{bmatrix}; \quad \begin{bmatrix} \underset{\sim}{Q}_2 \\ 1 \end{bmatrix} = D_1 \begin{bmatrix} \underset{\sim}{Q}_{20} \\ 1 \end{bmatrix} = D_1 \begin{bmatrix} 0 \\ 0 \\ 0 \\ 1 \end{bmatrix} = \begin{bmatrix} 0 \\ 0 \\ 0 \\ 1 \end{bmatrix}$$

$$\underset{\sim}{u}_2 \times \underset{\sim}{Q}_2 \underset{\sim}{H} = \underset{\sim}{u}_2 \times \underset{\sim}{H} = \begin{bmatrix} s_1 H_z \\ -c_1 H_z \\ c_1 H_y - s_1 H_x \end{bmatrix} = \begin{bmatrix} -b s_1 c_2 s_3 \\ b c_1 c_2 s_3 \\ b s_2 s_3 \end{bmatrix}$$

$$\underset{\sim}{u}_3 = R_1 R_2 \underset{\sim}{u}_{30} = R_1 R_2 \begin{bmatrix} 0 \\ 1 \\ 0 \end{bmatrix} = \begin{bmatrix} -s_1 c_2 \\ c_1 c_2 \\ s_2 \end{bmatrix}; \quad \begin{bmatrix} \underset{\sim}{Q}_3 \\ 1 \end{bmatrix} = D_1 D_2 \begin{bmatrix} a \\ 0 \\ 0 \\ 1 \end{bmatrix} = \begin{bmatrix} a c_1 \\ a s_1 \\ 0 \\ 1 \end{bmatrix}$$

$$\underset{\sim}{u}_3 \times \underset{\sim}{Q}_3 \underset{\sim}{H} = \underset{\sim}{u}_3 \times (\underset{\sim}{H} - \underset{\sim}{Q}_3) = \begin{bmatrix} -b c_1 s_3 - b s_1 s_2 c_3 \\ b c_1 s_2 c_3 - b s_1 s_3 \\ -b c_2 c_3 \end{bmatrix}$$

21. (continued)

$$J_2 = \begin{bmatrix} -(a+bc_3)s_1 - bc_1s_2s_3 & -bs_1c_2s_3 & -bc_1s_3 - bs_1s_2c_3 \\ (a+bc_3)c_1 - bs_1s_2s_3 & bc_1c_2s_3 & -bs_1s_3 + bc_1s_2c_3 \\ 0 & bs_2s_3 & -bc_2c_3 \end{bmatrix}$$

Same result is found as

$$J_2 = \nabla^t \underset{\sim}{H} = \begin{bmatrix} \frac{\partial H_x}{\partial \theta_1} & \frac{\partial H_x}{\partial \theta_2} & \frac{\partial H_x}{\partial \theta_3} \\ \frac{\partial H_y}{\partial \theta_1} & \frac{\partial H_y}{\partial \theta_2} & \frac{\partial H_y}{\partial \theta_3} \\ \frac{\partial H_z}{\partial \theta_1} & \frac{\partial H_z}{\partial \theta_2} & \frac{\partial H_z}{\partial \theta_3} \end{bmatrix}$$

$$\text{Det}\,|J_2| = \cdots = -ab^2 s_2 s_3^2 = -ab^2 \sin\theta_2 \sin^2\theta_3$$

23.

See next page for Problem 22

t	x	$\dot{x}$
1.0	5.000000	10.000000
1.1	6.105100	12.154000
1.2	7.441600	14.632000
1.3	9.043100	17.458000
1.4	10.945600	x

Adams Predictor: $x(1.4) = x(1.3) + \frac{0.1}{24}\left[55\,\dot{x}(1.3) - 59\,\dot{x}(1.2) + 37\,\dot{x}(1.1) - 9\,\dot{x}(1.0)\right]$

$= 10.945600$ (matches the exact value; error $\sim \left(\frac{d^5x}{dt^5}\right)(.1)^5$)

22.

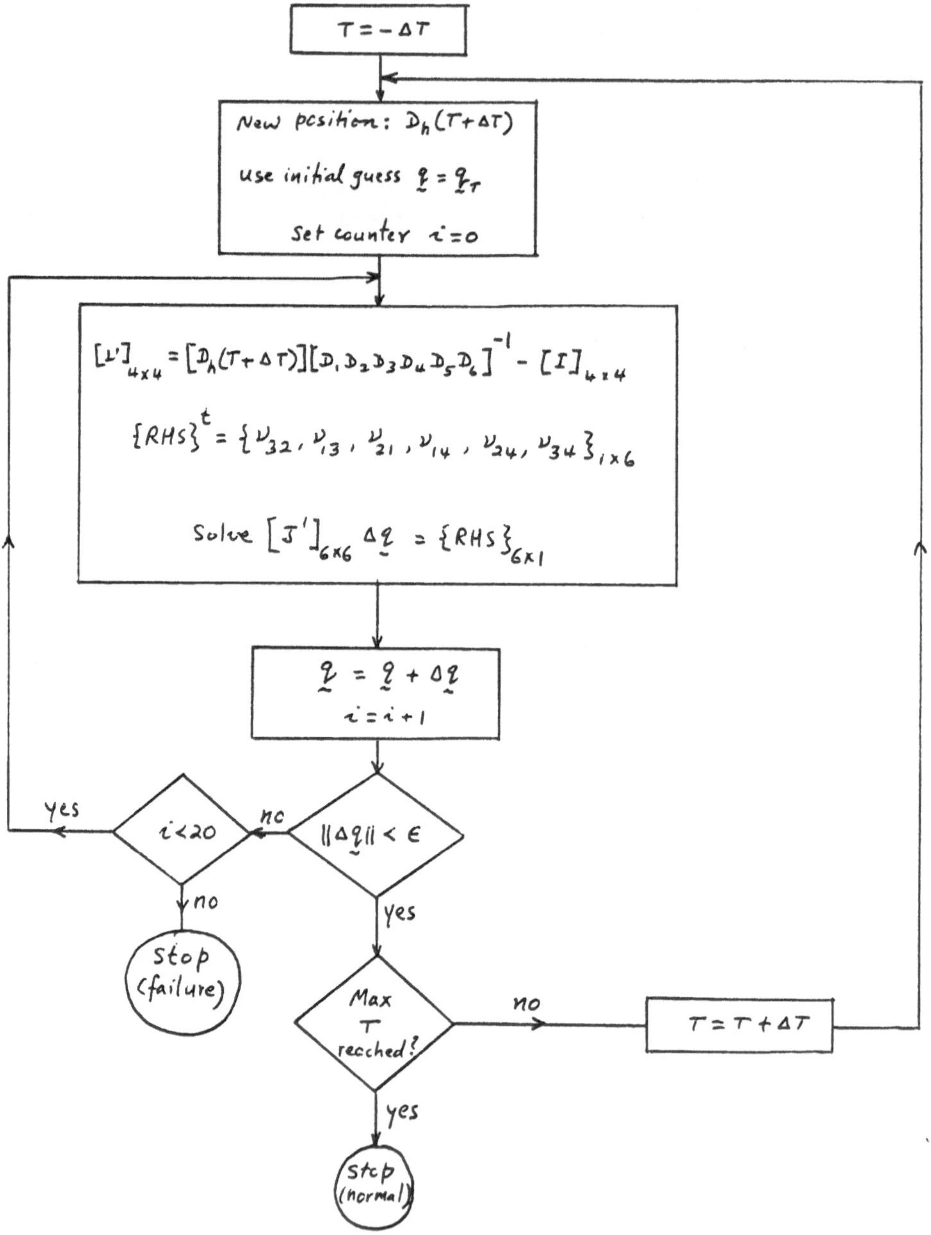
$T = -\Delta T$
New position: $D_h(T+\Delta T)$
use initial guess $\underset{\sim}{q} = \underset{\sim}{q}_T$
set counter $i=0$
$[L']_{4\times4} = [D_h(T+\Delta T)][D_1 D_2 D_3 D_4 D_5 D_6]^{-1} - [I]_{4\times4}$
$\{RHS\}^t = \{\nu_{32}, \nu_{13}, \nu_{21}, \nu_{14}, \nu_{24}, \nu_{34}\}_{1\times6}$
Solve $[J']_{6\times6} \Delta\underset{\sim}{q} = \{RHS\}_{6\times1}$
$\underset{\sim}{q} = \underset{\sim}{q} + \Delta\underset{\sim}{q}$
$i = i+1$
$\|\Delta\underset{\sim}{q}\| < \epsilon$
no
$i<20$
yes
no
Stop (failure)
yes
Max T reached?
no
$T = T + \Delta T$
yes
stop (normal)

CHAPTER 3

1.

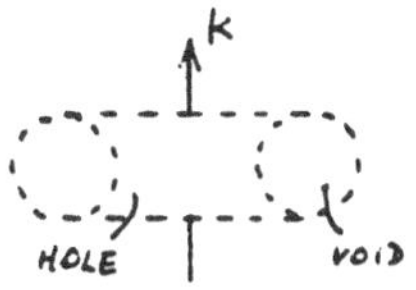

$W_k(P)$ is a torus

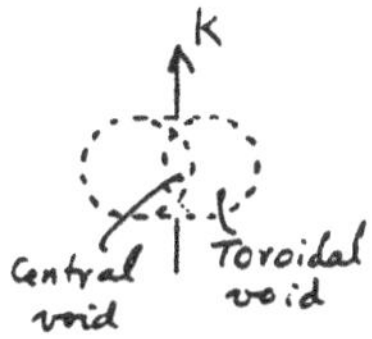

$W_k(P)$: sphere

2.

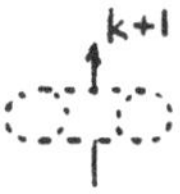

$W_{k+1}(P)$ is a donut

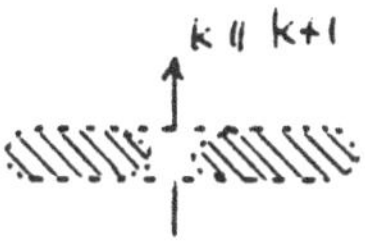

k is outside $W_{k+1}(P)$

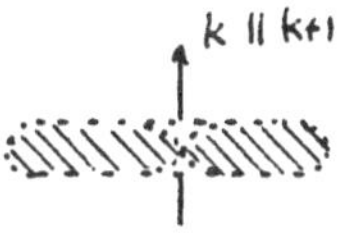

k cuts $W_{k+1}(P)$

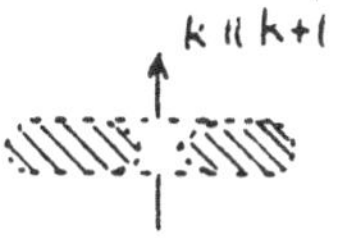

k passes through the hole of $W_{k+1}(P)$

3. Not to scale:

For $1' \leq S_3 \leq 5'$:

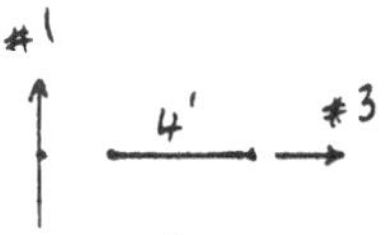

$W_3(H)$: line segment

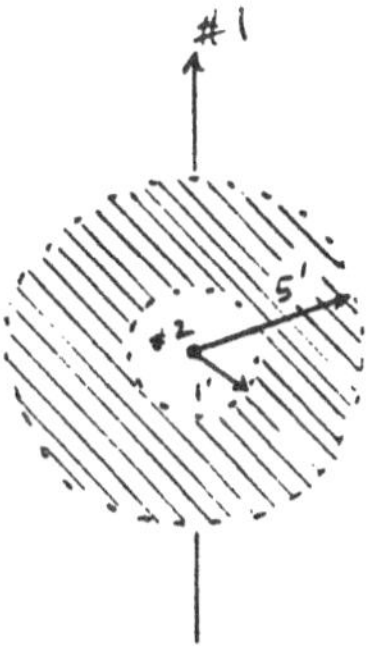

$W_2(H)$: disk with hole

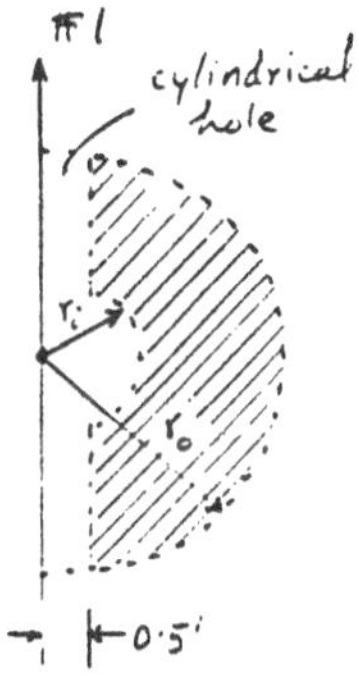

$r_i = \sqrt{1+.5^2}$, $r_o = \sqrt{5^2+.5^2}$

(spherical boundaries)

$W_1(H)$: spherical shell with cylindrical hole

For $0 \leq S_3 \leq 5'$:

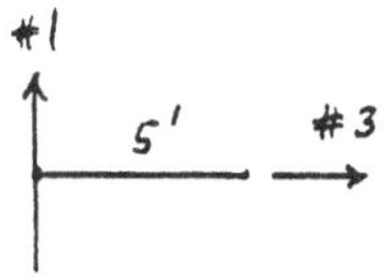

$W_3(H)$: line segment

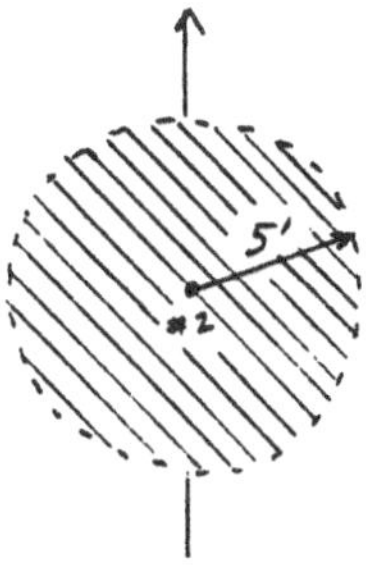

$W_2(H)$: solid disk

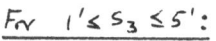

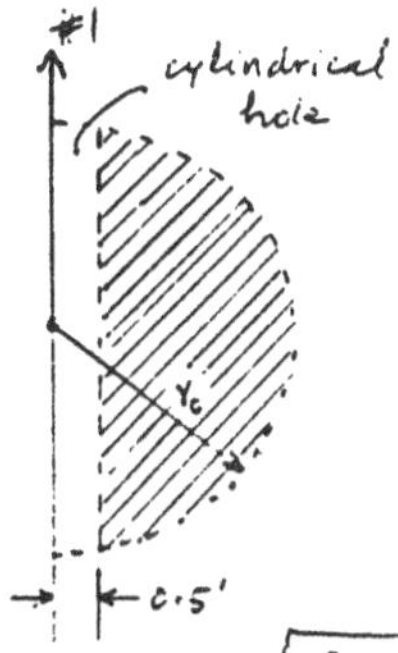

$r_o = \sqrt{5^2+.5^2}$

$W_1(H)$: sphere with a cylindrical hole

4.

$W_3(H)$: Circle

$w_2(H)$: torus without hole

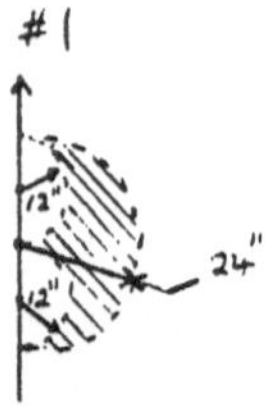

$w_1(H)$: sphere of radius 24" with two central voids of radius 12"

5. (a) $W_3(H)$ is a circle of radius 17"

$W_2(H)$ is a solid disk of radius 34"

$w_1(H)$ is a sphere of radius $OB = \sqrt{8^2 + 34^2}$, or $OB = 34.928$" with a cylindrical hole of radius $OA = 8$".

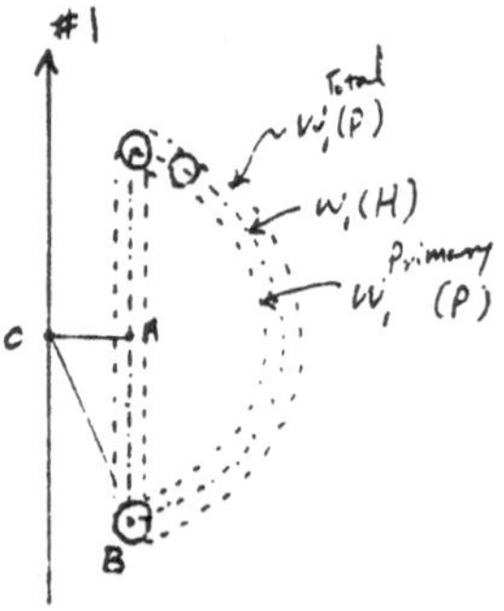

OA = 8", AB = 34", OB = 34.928"

(b) Ignoring corner effects & end effects,

$w_1^{\text{Primary}}(P)$ = sphere with 30.928" rad & 12" cyl. hole.

$w_1^{\text{Total}}(P)$ = sphere with 38.928" rad & 4" cyl. hole

These volumes can be found without using the Pappas theorem:

$$\text{Primary volume} \cong \frac{4\pi}{3}(30.928)^3 - \pi(12)^2 \cdot 2(30.928) \cong 95.94 \times 10^3 \text{ in}^3$$

$$\text{Total volume} \cong \frac{4\pi}{3}(38.928)^3 - \pi(4)^2 \cdot 2(38.928) \cong 243.19 \times 10^3 \text{ in}^3$$

Primary fraction ≅ 39.45%

6. $\alpha_{n-2} = 100^\circ, \alpha_{n-1} = 50^\circ, \alpha_n = 25^\circ$

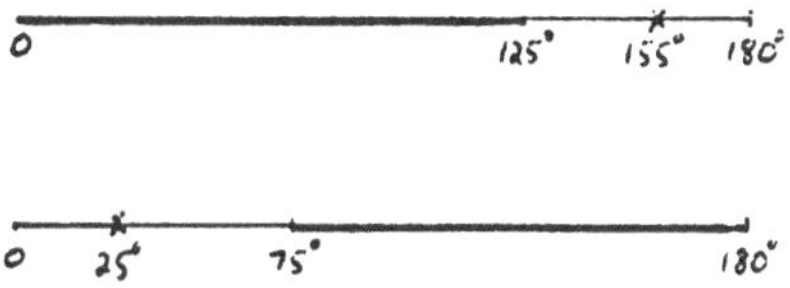

$|\alpha_f + \alpha_n - \pi| > |\alpha_{n-1} + \alpha_{n-2} - \pi|$

&

$|\alpha_f - \alpha_n| > |\alpha_{n-1} - \alpha_{n-2}|$

$\Rightarrow |\alpha_f - 155^\circ| > 30^\circ$

&

$|\alpha_f - 25^\circ| > 50^\circ$

$\Rightarrow \alpha_f < 125^\circ, \alpha_f > 185^\circ$

& $\alpha_f < -25^\circ, \alpha_f > 75^\circ$

Ignore values beyond $[0, 180^\circ]$:

$0 < \alpha_f < 125^\circ$ (see figure)

&

$75^\circ < \alpha_f < 180^\circ$ (see figure)

common range: $75^\circ < \alpha_f < 125^\circ$ for complete tool spin.

7. $\alpha_{n-2} = 100^\circ, \alpha_{n-1} = 85^\circ, \alpha_n = 80^\circ$

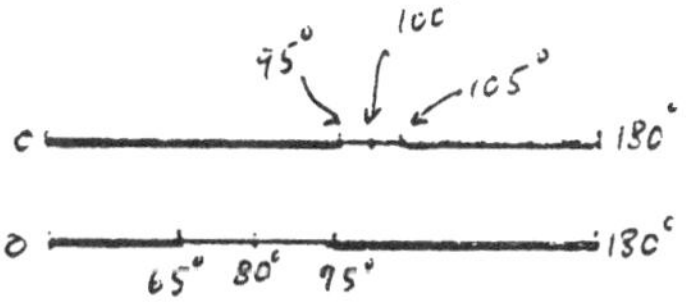

$\Rightarrow |\alpha_f - 100^\circ| > 5^\circ$

& $|\alpha_f - 80^\circ| > 15^\circ$

$\Rightarrow \alpha_f < 95^\circ, \alpha_f > 105^\circ$

& $\alpha_f < 65^\circ, \alpha_f > 95^\circ$

For range $[0, 180^\circ]$,

$0 < \alpha_f < 95^\circ$, $105^\circ < \alpha_f < 180^\circ$ (see figure)

& $0 < \alpha_f < 65^\circ$, $95^\circ < \alpha_f < 180^\circ$ (see figure)

Common ranges: $0 < \alpha_f < 65^\circ$ & $105^\circ < \alpha_f < 180^\circ$ for complete tool spin.

8. $\alpha_{n-2} = 120°, \alpha_{n-1} = 60°, \alpha_n = 15°$

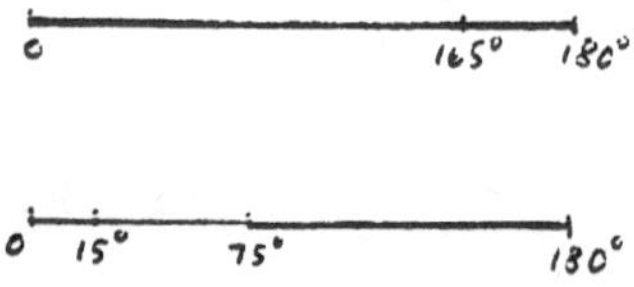

$\Rightarrow |\alpha_f - 165°| > 0°$

& $|\alpha_f - 15°| > 60°$

$\Rightarrow \alpha_f \neq 165°$

& $\alpha_f < -45°$, $\alpha_f > 75°$

For the range $[0, 180°]$,

$0 < \alpha_f < 165°$, $165° < \alpha_f < 180°$

& $75° < \alpha_f < 180°$

Common range: $75° < \alpha_f < 180°$, $\alpha_f \neq 165°$

1. $$\underset{\sim}{M}_G^{\text{Inertia}} = -[I_G]\underset{\sim}{\alpha} - [\Omega][I_G]\underset{\sim}{\omega} \qquad (4.22)$$

$$= -\begin{bmatrix} I_1 & 0 & 0 \\ 0 & I_2 & 0 \\ 0 & 0 & I_3 \end{bmatrix}\begin{bmatrix} \alpha_1 \\ \alpha_2 \\ \alpha_3 \end{bmatrix} - \begin{bmatrix} 0 & -\omega_3 & \omega_2 \\ \omega_3 & 0 & -\omega_1 \\ -\omega_2 & \omega_1 & 0 \end{bmatrix}\begin{bmatrix} I_1\omega_1 \\ I_2\omega_2 \\ I_3\omega_3 \end{bmatrix}$$

$$= \begin{bmatrix} -I_1\alpha_1 - (I_3 - I_2)\omega_2\omega_3 \\ -I_2\alpha_2 - (I_1 - I_3)\omega_1\omega_3 \\ -I_3\alpha_3 - (I_2 - I_1)\omega_1\omega_2 \end{bmatrix} \qquad (4.23)$$

2. $$(\underset{\sim}{F}^{ext})_{tb} = m\,(\underset{\sim}{a}_G)_{tb} \qquad (4.6)$$

Premultiply by $[R]^t$,

$$[R]^t(\underset{\sim}{F}^{ext})_{tb} = m\,[R]^t(\underset{\sim}{a}_G)_{tb} \quad ; \text{ next use } (4.27)$$

$$\Rightarrow (\underset{\sim}{F}^{ext})^* = m\,(\underset{\sim}{a}_G)^* \qquad (4.28)$$

$$(\underset{\sim}{M}_G^{ext})_{tb} = [I_G]_{tb}(\underset{\sim}{\alpha})_{tb} + [\Omega]_{tb}[I_G]_{tb}(\underset{\sim}{\omega})_{tb} \qquad (4.26)$$

Premultiply by $[R]^t$,

$$[R]^t(\underset{\sim}{M}_G^{ext})_{tb} = [R]^t[I_G]_{tb}\,(\underset{\sim}{\alpha})_{tb} + [R]^t[\Omega]_{tb}\,[I_G]_{tb}\,(\underset{\sim}{\omega})_{tb}$$

insert $[R][R]^t$

$$= [R]^t[I_G]_{tb}[R].[R]^t(\underset{\sim}{\alpha})_{tb}$$

$$+ [R]^t[\Omega]_{tb}[R].[R]^t[I_G]_{tb}[R].[R]^t(\underset{\sim}{\omega})_{tb}$$

using eq^n (4.27),

$$(\underset{\sim}{M}_G^{ext})^* = [I_{G0}](\underset{\sim}{\alpha})^* + [\Omega]^*[I_{G0}](\underset{\sim}{\omega})^* \qquad (4.29)$$

3. $R^L_{k+1} = R^L_k \, R(\theta_k, \underset{\sim}{u}_{ko})$ (4.31)

$\underset{\sim}{\omega}_{k+1} = \underset{\sim}{\omega}_k + \dot{\theta}_k \, \underset{\sim}{u}_k$ (4.33)

Note that $(\underset{\sim}{\omega}_k)^* = [R^L_k]^t \underset{\sim}{\omega}_k$ & $(\underset{\sim}{\omega}_{k+1})^* = [R^L_{k+1}]^t \underset{\sim}{\omega}_{k+1}$

Premultiply eq^n (4.33) by $[R^L_{k+1}]^t$:

$$[R^L_{k+1}]^t \underset{\sim}{\omega}_{k+1} = [R^L_{k+1}]\{\underset{\sim}{\omega}_k + \dot{\theta}_k \underset{\sim}{u}_k\}$$

$$= [R(\theta_k, \underset{\sim}{u}_{ko})]^t [R^L_k]^t \{\underset{\sim}{\omega}_k + \dot{\theta}_k \underset{\sim}{u}_k\}$$

$$= [R(\theta_k, \underset{\sim}{u}_{ko})]^t \{[R^L_k]^t \underset{\sim}{\omega}_k + \dot{\theta}_k [R^L_k]^t \underset{\sim}{u}_k\}$$

$$\Rightarrow (\underset{\sim}{\omega}_{k+1})^* = [R(\theta_k, \underset{\sim}{u}_{ko})]^t \{(\underset{\sim}{\omega}_k)^* + \dot{\theta}_k \underset{\sim}{u}_{ko}\} \qquad (4.35a)$$

Noting that $[R(\theta_k, \underset{\sim}{u}_{ko})]^t$ does not change $\underset{\sim}{u}_{ko}$,

$$(\underset{\sim}{\omega}_{k+1})^* = [R(\theta_k, \underset{\sim}{u}_{ko})]^t (\underset{\sim}{\omega}_k)^* + \dot{\theta}_k \underset{\sim}{u}_{ko} \qquad (4.35b)$$

Then, rearrange (4.35a),

$$(\underset{\sim}{\omega}_k)^* = [R(\theta_k, \underset{\sim}{u}_{ko})](\underset{\sim}{\omega}_{k+1})^* - \dot{\theta}_k \underset{\sim}{u}_{ko} \qquad (4.37)$$

4. $\underset{\sim}{\alpha}_{k+1} = \underset{\sim}{\alpha}_k + \ddot{\theta}_k \underset{\sim}{u}_k + \dot{\theta}_k \underset{\sim}{\omega}_k \times \underset{\sim}{u}_k$ (4.34)

Premultiply by $[R^L_{k+1}]^t$ & note that $R^L_{k+1} = R^L_k R(\theta_k, \underset{\sim}{u}_{ko})$,

$$[R^L_{k+1}]^t \underset{\sim}{\alpha}_{k+1} = [R(\theta_k, \underset{\sim}{u}_{ko})]^t \{[R^L_k]\underset{\sim}{\alpha}_k + \ddot{\theta}_k [R^L_k] \underset{\sim}{u}_k + \dot{\theta}_k [R^L_k](\underset{\sim}{\omega}_k \times \underset{\sim}{u}_k)\}$$

or $$(\underset{\sim}{\alpha}_{k+1})^* = [R(\theta_k, \underset{\sim}{u}_{ko})]^t \{(\underset{\sim}{\alpha}_k)^* + \ddot{\theta}_k \underset{\sim}{u}_{ko} + \dot{\theta}_k [R^L_k]\underset{\sim}{\omega}_k \times [R^L_k]\underset{\sim}{u}_k\}$$

$$= [R(\theta_k, \underset{\sim}{u}_{ko})]^t \{(\underset{\sim}{\alpha}_k)^* + \ddot{\theta}_k \underset{\sim}{u}_{ko} + \dot{\theta}_k (\underset{\sim}{\omega}_k)^* \times \underset{\sim}{u}_{ko}\}$$

$$= [R(\theta_k, \underset{\sim}{u}_{ko})]^t (\underset{\sim}{\alpha}_k)^* + \ddot{\theta}_k \underset{\sim}{u}_{ko} + \dot{\theta}_k [R(\theta_k, \underset{\sim}{u}_{ko})]^t (\underset{\sim}{\omega}_k)^* \times \underset{\sim}{u}_{ko} \qquad (4.36)$$

(note that $[R(\theta_k, \underset{\sim}{u}_{ko})]$ does not affect $\underset{\sim}{u}_{ko}$)

5. $$\underset{\sim}{v}_{p_k^k} = \underset{\sim}{v}_{G_k} + \underset{\sim}{\omega}_k \times \underset{\sim}{d}_k$$

Premultiply by $[R_k^L]^t$

$$[R_k^L]^t \underset{\sim}{v}_{p_k^k} = [R_k^L]^t \underset{\sim}{v}_{G_k} + [R_k^L]^t (\underset{\sim}{\omega}_k \times \underset{\sim}{d}_k)$$

$$\underset{\sim}{v}^*_{p_k^k} = \underset{\sim}{v}^*_{G_k} + [R_k^L]^t \underset{\sim}{\omega}_k \times [R_k^L]^t \underset{\sim}{d}_k$$

$$= \underset{\sim}{v}^*_{G_k} + \underset{\sim}{\omega}^*_k \times \underset{\sim}{d}^*_k$$

For a revolute joint,

$$\underset{\sim}{v}_{p_k^{k+1}} = \underset{\sim}{v}_{p_k^k}$$

Premultiply by $[R_{k+1}^L]^t$ & note that $R_{k+1}^L = R_k^L \cdot R(\theta_k, \underset{\sim}{u}_{ko})$,

$$[R_{k+1}^L]^t \underset{\sim}{v}_{p_k^{k+1}} = [R(\theta_k, \underset{\sim}{u}_{ko})]^t \cdot [R_k^L]^t \underset{\sim}{v}_{p_k^k}$$

or, $$\underset{\sim}{v}^*_{p_k^{k+1}} = [R(\theta_k, \underset{\sim}{u}_{ko})]^t \underset{\sim}{v}^*_{p_k^k}$$

For a prismatic joint,

$$\underset{\sim}{v}_{p_k^{k+1}} = \underset{\sim}{v}_{p_k^k} + \dot{s}_k \underset{\sim}{u}_k \qquad \text{(see 4.3)}$$

Premultiply by $[R_{k+1}^L]^t$ and simplify,

$$\underset{\sim}{v}^*_{p_k^{k+1}} = \underbrace{[R(\theta_k, \underset{\sim}{u}_{ko})]^t}_{[I]} (\underset{\sim}{v}^*_{p_k^k} + \dot{s}_k \underset{\sim}{u}_{ko}), \quad \text{but } \theta_k = 0$$

$$= \underset{\sim}{v}^*_{p_k^k} + \dot{\underset{\sim}{s}}_k \underset{\sim}{u}_{ko}$$

5. (continued)

Finally, $\underset{\sim}{v}_{G_{k+1}} = \underset{\sim}{v}_{P_k^{k+1}} + \underset{\sim}{\omega}_{k+1} \times \underset{\sim}{\varsigma}_{k+1}$

Premultiply by $[R_{k+1}^L]^t$ & simplify,

$$\underset{\sim}{v}^*_{G_{k+1}} = \underset{\sim}{v}^*_{P_k^{k+1}} + \underset{\sim}{\omega}^*_{k+1} \times \underset{\sim}{\varsigma}^*_{k+1}$$

6. Details are similar to Prob. 5.

$$\underset{\sim}{a}_{P_k^k} = \underset{\sim}{a}_{G_k} + \underset{\sim}{\omega}_k \times (\underset{\sim}{\omega}_k \times \underset{\sim}{d}_k) + \underset{\sim}{\alpha}_k \times \underset{\sim}{d}_k$$

Premultiply by $[R_k^L]^t$ and simplify,

$$\underset{\sim}{a}^*_{P_k^k} = \underset{\sim}{a}^*_{G_k} + [R_k^L]^t \underset{\sim}{\omega}_k \times \left([R_k^L]^t \underset{\sim}{\omega}_k \times [R_k^L]^t \underset{\sim}{d}_k\right) + [R_k^L]^t \underset{\sim}{\alpha}_k \times [R_k^L]^t \underset{\sim}{d}_k$$

$$= \underset{\sim}{a}^*_{G_k} + \underset{\sim}{\omega}^*_k \times (\underset{\sim}{\omega}^*_k \times \underset{\sim}{d}^*_k) + \underset{\sim}{\alpha}^*_k \times \underset{\sim}{d}^*_k$$

For a revolute joint,

$$\underset{\sim}{a}_{P_k^{k+1}} = \underset{\sim}{a}_{P_k^k}$$

Premultiply by $[R_{k+1}^L]^t$ and noting $R_{k+1}^L = R_k^L \cdot R(\theta_k, \underset{\sim}{y}_{ko})$,

$$\underset{\sim}{a}^*_{P_k^{k+1}} = [R(\theta_k, \underset{\sim}{y}_{ko})]^t \underset{\sim}{a}^*_{P_k^k}$$

6. (continued)

For a prismatic joint,

$$\underset{\sim}{a}_{p_k^{k+1}} = \underset{\sim}{a}_{p_k^k} + \ddot{s}_k \underset{\sim}{u}_k + 2\underset{\sim}{\omega}_k \times (\dot{s}_k \underset{\sim}{u}_k) \qquad \text{(See 4.4)}$$

Premultiply by $[R_{k+1}^L]^t$ & simplify,

$$\underset{\sim}{a}^*_{p_k^{k+1}} = [R(\theta_k, \underset{\sim}{u}_{ko})]^t \left(\underset{\sim}{a}^*_{p_k^k} + \ddot{s}_k \underset{\sim}{u}_{ko} + 2\dot{s}_k \underset{\sim}{\omega}^*_k \times \underset{\sim}{u}_{ko} \right)$$

But, for prismatic joint k, $\theta_k = 0$ and $R(\theta_k, \underset{\sim}{u}_{ko}) = I$.

$$\Rightarrow \underset{\sim}{a}^*_{p_k^{k+1}} = \underset{\sim}{a}^*_{p_k^k} + \ddot{s}_k \underset{\sim}{u}_{ko} + 2\dot{s}_k \underset{\sim}{\omega}^*_k \times \underset{\sim}{u}_{ko}$$

$$\underset{\sim}{a}_{G_{k+1}} = \underset{\sim}{a}_{p_k^{k+1}} + \underset{\sim}{\omega}_{k+1} \times (\underset{\sim}{\omega}_{k+1} \times \underset{\sim}{c}_{k+1}) + \underset{\sim}{\alpha}_{k+1} \times \underset{\sim}{c}_{k+1}$$

Premultiply by $[R_{k+1}^L]^t$ & simplify

$$\underset{\sim}{a}^*_{G_{k+1}} = \underset{\sim}{a}^*_{p_k^{k+1}} + \underset{\sim}{\omega}^*_{k+1} \times (\underset{\sim}{\omega}^*_{k+1} \times \underset{\sim}{c}^*_{k+1}) + \underset{\sim}{\alpha}^*_{k+1} \times \underset{\sim}{c}^*_{k+1}$$

7. For revolute jointed manipulator,

$$(\underset{\sim}{\omega}_{k+1})^* = [R(\theta_k, \underset{\sim}{u}_{k0})]^t (\underset{\sim}{\omega}_k)^* + \dot{\theta}_k \underset{\sim}{u}_{k0} \qquad (4.35)$$

$$(\underset{\sim}{v}_{G_{k+1}})^* = (\underset{\sim}{v}_{p_k^{k+1}})^* + (\underset{\sim}{\omega}_{k+1})^* \times \underset{\sim}{s}_{k+1,0} \qquad \text{(Prob. 5 \& pg. 144)}$$

$$T = \frac{1}{2}\sum_k m_k (\underset{\sim}{v}_{G_k})^{*t} (\underset{\sim}{v}_{G_k})^* + \frac{1}{2}\sum_k (\underset{\sim}{\omega}_k)^{*t} [I_{G_{k,0}}] (\underset{\sim}{\omega}_k)^* \qquad (4.50)$$

$$V = \sum_k m_k g (z_{G_k} - z_{G_{k,0}})$$

(a) $\underset{\sim}{\omega}_1^* = 0$

$$\underset{\sim}{\omega}_2^* = \cancel{[R(\theta_1, \underset{\sim}{u}_{10})]^t \underset{\sim}{\omega}_1^*}^{\,0} + \dot{\theta}_1 \underset{\sim}{u}_{10} = \dot{\theta}_1 \begin{bmatrix} 0 \\ 0 \\ 1 \end{bmatrix}$$

$$\underset{\sim}{\omega}_3^* = [R(\theta_2, \underset{\sim}{u}_{20})]^t \underset{\sim}{\omega}_2^* + \dot{\theta}_2 \underset{\sim}{u}_{20}$$

$$= \begin{bmatrix} c\theta_2 & 0 & -s\theta_2 \\ 0 & 1 & 0 \\ s\theta_2 & 0 & c\theta_2 \end{bmatrix} \begin{bmatrix} 0 \\ 0 \\ \dot{\theta}_1 \end{bmatrix} + \begin{bmatrix} 0 \\ \dot{\theta}_2 \\ 0 \end{bmatrix} = \begin{bmatrix} -\dot{\theta}_1 s\theta_2 \\ \dot{\theta}_2 \\ \dot{\theta}_1 c\theta_2 \end{bmatrix}$$

$\underset{\sim}{v}_{G_1}^* = \underset{\sim}{v}_{G_2}^* = 0$ (trivial)

$$\underset{\sim}{v}_{G_3}^* = \cancel{\underset{\sim}{v}_{p_2^3}^*}^{\,0} + \underset{\sim}{\omega}_3^* \times \underset{\sim}{s}_{30} = \begin{vmatrix} \underset{\sim}{i} & \underset{\sim}{j} & \underset{\sim}{k} \\ -\dot{\theta}_1 s\theta_2 & \dot{\theta}_2 & \dot{\theta}_1 c\theta_2 \\ l & 0 & 0 \end{vmatrix} = \begin{bmatrix} 0 \\ l\dot{\theta}_1 c\theta_2 \\ -l\dot{\theta}_2 \end{bmatrix}$$

(b)

$$\underset{\sim}{s}_3 = D_1 D_2 \underset{\sim}{s}_{30} = \begin{bmatrix} c\theta_1 & -s\theta_1 & 0 & 0 \\ s\theta_1 & c\theta_1 & 0 & 0 \\ 0 & 0 & 1 & 0 \\ 0 & 0 & 0 & 1 \end{bmatrix} \begin{bmatrix} c\theta_2 & 0 & s\theta_2 & 0 \\ 0 & 1 & 0 & 0 \\ -s\theta_2 & 0 & c\theta_2 & 0 \\ 0 & 0 & 0 & 1 \end{bmatrix} \begin{bmatrix} l \\ 0 \\ 0 \\ 1 \end{bmatrix}$$

$$= \begin{bmatrix} l c\theta_1 c\theta_2 \\ l s\theta_1 c\theta_2 \\ -l s\theta_2 \\ 1 \end{bmatrix} \equiv \begin{bmatrix} x_{G3} \\ y_{G3} \\ z_{G3} \\ 1 \end{bmatrix}$$

7. (continued)

$$T = \frac{1}{2} m_3 \ell^2 (\dot{\theta}_1^2 \cos^2\theta_2 + \dot{\theta}_2^2) + \frac{1}{2}(I_x \dot{\theta}_1^2 \sin^2\theta_2 + \dot{\theta}_2^2 I_y + I_z \dot{\theta}_1^2 \cos^2\theta_2)$$

$$V = -m_3 g \ell \sin\theta_2$$

$$L = T - V$$

$$= \frac{1}{2} m_3 \ell^2 (\dot{\theta}_1^2 \cos^2\theta_2 + \dot{\theta}_2^2) + \frac{1}{2}(I_x \dot{\theta}_1^2 \sin^2\theta_2 + I_y \dot{\theta}_2^2 + I_z \dot{\theta}_1^2 \cos^2\theta_2) + m_3 g \ell \sin\theta_2$$

$$\frac{\partial L}{\partial \dot{\theta}_1} = m_3 \ell^2 \dot{\theta}_1 \cos^2\theta_2 + I_x \dot{\theta}_1 \sin^2\theta_2 + I_z \dot{\theta}_1 \cos^2\theta_2$$

$$= \dot{\theta}_1 \left[(I_z + m\ell^2)\cos^2\theta_2 + I_x \sin^2\theta_2 \right]$$

$$\frac{\partial L}{\partial \dot{\theta}_2} = m_3 \ell^2 \dot{\theta}_2 + I_y \dot{\theta}_2$$

$$= \dot{\theta}_2 (I_y + m\ell^2)$$

$$\frac{d}{dt}\left(\frac{\partial L}{\partial \dot{\theta}_1}\right) = \ddot{\theta}_1 \left[(I_z + m\ell^2)\cos^2\theta_2 + I_x \sin^2\theta_2 \right] + 2\dot{\theta}_1 \dot{\theta}_2 (I_x - I_z - m_3 \ell^2)\sin\theta_2 \cos\theta_2$$

$$\frac{d}{dt}\left(\frac{\partial L}{\partial \dot{\theta}_2}\right) = \ddot{\theta}_2 (I_y + m_3 \ell^2)$$

$$\frac{\partial L}{\partial \theta_1} = 0$$

$$\frac{\partial L}{\partial \theta_2} = (I_x - I_z - m_3 \ell^2)\dot{\theta}_1^2 \sin\theta_2 \cos\theta_2 + m_3 g \ell \cos\theta_2$$

substitute into

$$\frac{d}{dt}\left(\frac{\partial L}{\partial \dot{\theta}_i}\right) - \frac{\partial L}{\partial \theta_i} = \tau_i \quad , i = 1, 2$$

7. (continued)

$$\left[(I_z + m_3 l^2)\cos^2\theta_2 + I_x \sin^2\theta_2\right]\ddot{\theta}_1 + \left[2(I_x - I_z - m_3 l^2)\sin\theta_2\cos\theta_2\right]\dot{\theta}_1\dot{\theta}_2 = \tau_1$$

$$(I_y + m_3 l^2)\ddot{\theta}_2 - \left[(I_x - I_z - m_3 l^2)\sin\theta_2\cos\theta_2\right]\dot{\theta}_1^2 - m_3 g l \cos\theta_2 = \tau_2$$

(c)

$$[H(\underset{\sim}{\theta})] = \begin{bmatrix} (I_z + m_3 l^2)\cos^2\theta_2 + I_x \sin^2\theta_2 & 0 \\ 0 & I_y + m_3 l^2 \end{bmatrix}_{2\times 2}$$

$$\underset{\sim}{c}(\underset{\sim}{\theta}, \dot{\underset{\sim}{\theta}}) = \begin{bmatrix} 2\dot{\theta}_1\dot{\theta}_2 (I_x - I_z - m_3 l^2)\sin\theta_2\cos\theta_2 \\ -\dot{\theta}_1^2 (I_x - I_z - m_3 l^2)\sin\theta_2\cos\theta_2 \end{bmatrix}_{2\times 1}$$

$$\underset{\sim}{g}(\underset{\sim}{\theta}) = \begin{bmatrix} 0 \\ -m_3 g l \cos\theta_2 \end{bmatrix}$$

$$\underset{\sim}{f} = \begin{bmatrix} \tau_1 \\ \tau_2 \end{bmatrix}$$

Control law:

$$\underset{\sim}{f}_{CL} = [H(\underset{\sim}{\theta}_a)]\left(\ddot{\underset{\sim}{\theta}}_d + [C_p]\underset{\sim}{\epsilon} + [C_v]\dot{\underset{\sim}{\epsilon}}\right) + \underset{\sim}{c}(\underset{\sim}{\theta}_a, \dot{\underset{\sim}{\theta}}_a) + \underset{\sim}{g}(\underset{\sim}{\theta}_a)$$

It will lead to error dynamics: $\ddot{\underset{\sim}{\epsilon}} + [C_v]\dot{\underset{\sim}{\epsilon}} + [C_p]\underset{\sim}{\epsilon} = \underset{\sim}{0}$

8. $I_c \ddot{\theta} + 0.1\, mL^2 \dot{\theta}^2 + 0.5\, mgL\cos\theta = \tau$

$H = I_o$

$c = 0.1\, mL^2 \dot{\theta}^2$

$g = 0.5\, mgL\cos\theta$

$\epsilon = \theta_d - \theta_a$

$\tau_{CL} = I_o(\ddot{\theta}_d + C_v \dot{\epsilon} + C_p \epsilon) + 0.1\, mL^2 \dot{\theta}_a^2 + 0.5\, mgL\cos\theta_a$

For $C_p = 225 \ \text{sec}^{-2}$, $C_v = 2\sqrt{C_p} = 30 \ \text{sec}^{-1}$

9. $(10\cos^2\theta_2)\ddot{\theta}_1 + (10\sin 2\theta_2)\dot{\theta}_1\dot{\theta}_2 = \tau_1$

$10\ddot{\theta}_2 + (5\sin 2\theta_2)\dot{\theta}_1^2 - 50\cos\theta_2 = \tau_2$

$$[H(\underset{\sim}{\theta})] = \begin{bmatrix} 10\cos^2\theta_2 & 0 \\ 0 & 10 \end{bmatrix}$$

$$\underset{\sim}{c}(\underset{\sim}{\theta}, \dot{\underset{\sim}{\theta}}) = \begin{bmatrix} (10\sin 2\theta_2)\dot{\theta}_1\dot{\theta}_2 \\ (5\sin 2\theta_2)\dot{\theta}_1^2 \end{bmatrix}$$

$$\underset{\sim}{g}(\underset{\sim}{\theta}) = \begin{bmatrix} 0 \\ -50\cos\theta_2 \end{bmatrix}$$

$$\underset{\sim}{\epsilon} = \underset{\sim}{\theta}_d - \underset{\sim}{\theta}_a = \begin{bmatrix} 29^\circ \\ 46^\circ \end{bmatrix} - \begin{bmatrix} 30^\circ \\ 45^\circ \end{bmatrix} = \begin{bmatrix} -1^\circ \\ 1^\circ \end{bmatrix} = \begin{bmatrix} -0.0175 \\ 0.0175 \end{bmatrix} \text{ rad}$$

$$\dot{\underset{\sim}{\epsilon}} = \dot{\underset{\sim}{\theta}}_d - \dot{\underset{\sim}{\theta}}_a = \begin{bmatrix} 2.0 \\ 5.5 \end{bmatrix} - \begin{bmatrix} 2.1 \\ 5.7 \end{bmatrix} = \begin{bmatrix} -.1 \\ -.2 \end{bmatrix} \text{ rad/s}$$

9. (continued)

$$\ddot{\underset{\sim}{Q}}_d = \begin{bmatrix} 10 \\ 15 \end{bmatrix} \text{ rad/s}^2$$

$$[C_v] = \begin{bmatrix} 20 & 0 \\ 0 & 20 \end{bmatrix} \text{ 1/s}$$

$$[C_p] = \begin{bmatrix} 100 & 0 \\ 0 & 100 \end{bmatrix} \text{ 1/s}^2$$

$$\underset{\sim}{\tau}_{CL} = [H(\underset{\sim}{Q}_a)]\left(\ddot{\underset{\sim}{Q}}_d + [C_v]\dot{\epsilon} + [C_p]\epsilon\right) + \underset{\sim}{c}(\underset{\sim}{Q}_a, \dot{\underset{\sim}{Q}}_a) + \underset{\sim}{g}(\underset{\sim}{Q}_a);$$

$$= \begin{bmatrix} 5 & 0 \\ 0 & 10 \end{bmatrix}\left(\begin{bmatrix} 10 \\ 15 \end{bmatrix} + \begin{bmatrix} -2 \\ -4 \end{bmatrix} + \begin{bmatrix} -1.75 \\ 1.75 \end{bmatrix}\right) + \begin{bmatrix} 119.7 \\ -13.3 \end{bmatrix}$$

$$= \begin{bmatrix} 150.95 \\ 113.70 \end{bmatrix}$$